SpringerBriefs in Applied Sciences and Technology

For further volumes:
http://www.springer.com/series/8884

Gottfried Spelsberg-Korspeter

Robust Structural Design against Self-Excited Vibrations

 Springer

Gottfried Spelsberg-Korspeter
Dynamics and Vibrations Group
TU Darmstadt
Darmstadt
Germany

ISSN 2191-530X   ISSN 2191-5318 (electronic)
ISBN 978-3-642-36551-5   ISBN 978-3-642-36552-2 (eBook)
DOI 10.1007/978-3-642-36552-2
Springer Heidelberg New York Dordrecht London

Library of Congress Control Number: 2013932334

Printed on acid-free paper

Springer is part of Springer Science+Business Media (www.springer.com)

Contents

Chapter 1
Introduction

Almost every engineering problem contains an optimization problem. Engineering structures are designed with respect to cost, weight, efficiency and many more objectives. Even if we limit our view to the vibrational behavior of machines and structures there are various goals to optimize for and several ambiguities.

In many technical applications structural vibrations are unwanted. On the other hand, as a rule of thumb, structures are designed to weigh less and less, which makes them even more accessible to unwanted vibrations. In several cases the source and frequency of the vibrations can be identified properly so that it is rather clear how to prevent the structure from unwanted motion. If we consider for example a slightly unbalanced rotor in a machine, this will result in a harmonic excitation of the structure. In case the excitation of a structure is more or less well known, it is easy to find countermeasures. For the problem of known harmonic excitation there exist many methods to design vibration absorbers which also tolerate slight modifications in frequency. If the excitation level is low and not permanent, often the introduction of damping is sufficient to prevent vibrations. Especially for linear vibrations there are many approaches and criteria on how to choose the design parameters for these cases [49].

The suppression of vibrations is more difficult if the excitation is not known exactly or difficult to quantify. This can occur due to the lack of predictability as for example in wind excited vibrations or due to the complexity of the structure. Even if the excitation is harmonic its frequency can vary over a wide range. Most technical applications are so complex that a robust dynamical model is hard to develop. If we think for example of a car, there are many materials involved and complex contact configurations occur. It is therefore impossible to develop fine models. On the other hand the goal is to have a quiet structure. Therefore it is important to design components robust against structure born sound.

A major limitation of the above mentioned optimization measures is that they require the structure to be stable, i.e. the occurrence of self-excited vibrations is ruled out a priory. However, self-excited vibrations play an important role in machinery where parts are moving and frictional contacts occur. The existing models for the

G. Spelsberg-Korspeter, *Robust Structural Design against Self-Excited Vibrations*, SpringerBriefs in Applied Sciences and Technology, DOI: 10.1007/978-3-642-36552-2_1, © The Author(s) 2013

excitation mechanisms are usually very crude, since the detailed processes occurring in contact areas between mechanical parts are practically impossible to describe and parameterize on a detailed level.

Practically oriented vibration engineers have developed several construction guidelines in order to avoid vibration problems mainly focusing on acoustical aspects. One of these rules is that unnecessary symmetries in a structure should be avoided. Mathematically this can be expressed by stating that multiple eigenfrequencies in a structure should be avoided. The goal of this book is to validate this construction guideline from a modeling point of view and to develop approaches to implement it in critical parts. It will be shown that in a wide class of configurations, the systematic avoidance of multiple eigenvalues prevents rotors in frictional contact from squealing. Examples for the systems within this category are disk and drum brakes, clutches, saws and paper calenders.

For brake squeal an overview about the topic can be found in the survey papers [27, 34, 59], for squealing clutches we mention [7–9]. The first step towards suppression of unwanted vibrations is to understand their origin at least to an extent that countermeasures can be taken. Especially for the problem of brake squeal, many recent papers consider the excitation mechanism (see e.g. [24, 50, 58, 70, 77, 91]), and, despite differences in modeling agree, that it is caused by sliding friction. While the excitation mechanism for brake squeal seems to be well understood, for the problem of paper calendering this is not the case [11, 87]. However, the paper of BROMMUNDT [2] indicates that sliding friction is a possible excitation mechanism for this application as well.

In the literature, in addition to sliding friction, also the effect of negative damping induced by a falling friction characteristic, stick slip and sprag slip effects [81] and nonlinear saturation effects [1] are proposed as excitation mechanisms [26, 63], which however do not play a role in the systems addressed in this book.

For the suppression of self-excited vibrations active, semi-active and passive measures have been suggested. Usually active measures are expensive and less reliable, since states of the system have to be measured and actuators have to be operated. Nevertheless they have been successfully used to suppress brake squeal in the laboratory [5, 12, 21, 63, 92]. Semi-active measures have for example been developed in [36, 53] using piezo ceramics attached to the brake pads.

An approach for the passive assignment of poles in nonconservative systems has been suggested by OUYANG [57]. It is noted however that the assignment of real parts of eigenvalues is much harder than the assignment of eigenfrequencies. The assignment of eigenfrequencies through passive modifications has been successfully done for example in [3, 39, 51, 52], the objective being an avoidance of resonances. However, in practice this strategy is often ruled out because eigenfrequencies change with operation speeds of a machine due to nonlinear characteristics.

Whereas most of the approaches mentioned above rely on the active and semi-active increase of dissipation, we want to influence mass and stiffness parameters to make a structure robust against self-excited vibrations. We will prove that structures with rotational symmetry are more sensitive to self-excited vibrations than systems without rotational symmetry, in agreement with the above mentioned design rule that multiple eigenfrequencies are to be avoided. A first observation in this direction was made in [47, 55], where it was noted that at least two modes of similar order in terms of eigenfrequency of the brake disk were necessary to capture squeal in pin-on-disk models.

The approach of braking symmetries in order to prevent squeal seems to have been investigated first experimentally by NISHIWAKI [54] for disk brakes and by LANG [42] for drum brakes. In a more recent approach, Fieldhouse and coworkers developed modified disk brakes in which problematic modes were hindered by structural modifications. The results presented in [9, 10] include experimental verification of the concept. In the above mentioned papers however no explanation for the stabilizing effect of the modifications is given and the excitation mechanism for squeal is not included in the calculations. Evidence that the mechanism behind the suppression of squeal is not understood in the community can also be found in [46], where modifications on a simplified disk brake are presented as a new approach towards the avoidance of brake squeal without giving a model based explanation of the finding.

In the context of linear systems with follower forces, it is well known that the splitting of eigenvalues of the stiffness matrix has a stabilizing effect, see eg. [26] and many others. The equations usually studied for these types of systems are of the form

$$M\ddot{q} + (K + N)q = 0, \tag{1.1}$$

with M, K constant, symmetric and positive definite and $N^T = -N$ constant, skew symmetric. These systems have also been studied in presence of damping which yields an interesting and complicated stability behavior of the trivial solution, often referred to as destabilization paradox [35]. Also for the more general case of a system of the type

$$M\ddot{q} + (D + G)\dot{q} + (K + N)q = 0, \tag{1.2}$$

with $D^T = D$ positive definite and $G^T = -G$ many results have been obtained, relying on the structure of the matrices of the second order system, see eg. [48]. In several technical systems it is, however, not possible to obtain second order equations of motion with constant coefficients even for the linearized case. Examples of this kind are asymmetric rotors with spatially fixed contacts. Intuitive examples of this kind are studied in Chap. 4. One goal of this book is therefore to generalize some of the results for (1.1) and (1.2) to linear conservative equations of motion with time-periodic perturbations. We will study equations of the type

$$M\ddot{q} + \Delta D(t)\dot{q} + (K + \Delta K(t))q = 0, \tag{1.3}$$
$$\Delta D(t) = \Delta D(t + T), \qquad \Delta K(t) = \Delta K(t + T),$$

where as before M, K are constant, symmetric and positive definite and $\Delta D(t)$, $\Delta K(t)$ are time periodic but have arbitrary structure. In applications the perturbation matrices $\Delta D(t)$ and $\Delta K(t)$ often arise due to contact forces.

The organization of the book is as follows. In the next chapter we explain the origination and the scope of the systems addressed in this book. Using analytic perturbation theory, we show that splitting of eigenvalues, which is achieved by means of disturbing rotational symmetry stabilizes the system and gives a robust design against self-excited vibrations. From this it follows that the design goal of splitting multiple frequencies is an appropriate objective for a structural optimization.

In Chap. 3 the underlying optimization problem will be motivated mechanically and the mathematical properties are investigated. This yields a justification for the optimization approaches chosen in later chapters.

Chapters 4 and 5 give simple examples of systems for which the developments are relevant and help to understand the structure of the underlying design problems.

In Chap. 6, as an engineering example, a disk brake is optimized with respect to its spectrum using different approaches. Afterwards an experimental verification of the theory is given.

Finally, in Chap. 7, the effect of nonlinearities on the occurrence of self-excited vibrations and their avoidance is investigated. It will be shown that the design rule of splitting eigenvalues of the linearized system is also beneficial in the nonlinear regime.

This book presents results obtained in the project DFG SP 1198/3-1 funded by the German Research Foundation and is to a large extend based on the research papers [72–76, 78, 80]. The text has been accepted as a Habilitation thesis for the field of Mechanics at the TU Darmstadt.

Chapter 2
Perturbation of a Linear Conservative System by Periodic Parametric Excitation

2.1 Motivation

Usually vibrations of machines and structures are defined about a certain state of operation. If the structure operates as planned, the deviations from this reference state are small, so that a linearized analysis of the corresponding equations of motion is meaningful and sufficient from a technical point of view. Usually one has rather good information about mass and stiffness properties of engineering structures, whereas other phenomena, such as contact conditions and damping, are hard to describe. The wide success of mechanical modeling relies on the fact that usually mass and stiffness properties dominate the vibrational behavior of mechanical structures. Mathematically this can be captured in terms of a perturbation problem

$$(M + \Delta M(t))\ddot{q} + \Delta D(t)\dot{q} + (K + \Delta K(t))q = F(t), \qquad (2.1)$$

where the mass matrix M and the stiffness matrix K describe the well known structural behavior of the parts with respect to their generalized coordinates, whereas the matrices $\Delta M(t)$, $\Delta D(t)$, $\Delta K(t)$ and the force vector $F(t)$ are perturbations, originating from small variable inertia, contacts, dissipation and unknown external forcing. For rigid body systems, the generalized coordinates can always be defined such that the linearized equations of motion have the form (2.1). In many cases, the perturbation matrices will either be constant or have periodic coefficients, as is the case for example in rotors with spatially fixed contacts. Since the perturbations originate from physical processes which are hard to model and to parameterize, they are impossible to predict accurately. Therefore a design goal must be to lay out the structure in such a way that the perturbations, which are only known to a little extend, do not cause significant vibrations. We will show in this chapter that it is a good strategy to design the system such that no multiple eigenfrequencies occur in the unperturbed problem. Usually for the consideration of (2.1) one divides the analysis into free vibrations and forced vibrations, the sum of which yield the general solution. However, the occurrence of the external forces $F(t)$ depends on the

G. Spelsberg-Korspeter, *Robust Structural Design against Self-Excited Vibrations*, SpringerBriefs in Applied Sciences and Technology, DOI: 10.1007/978-3-642-36552-2_2, © The Author(s) 2013

system boundary of the model. If one draws the boundary such that every forcing is generated within the model, only free vibrations occur. As an example consider the one degree of freedom (dof) oscillator

$$\ddot{x} + \omega_0^2 x = f \cos \Omega t. \tag{2.2}$$

By extending the system boundary such that the forcing is generated by another oscillator

$$\ddot{y} + \Omega^2 y = 0, \tag{2.3a}$$

$$\ddot{x} + \omega_0^2 x - \varepsilon y = 0, \tag{2.3b}$$

which is asymmetrically coupled to our original oscillator, the system can be modeled as a homogeneous problem. It is clear that asymmetric couplings are always possible, for example due to frictional contacts. Equation (2.3) is a special case of (2.1) with $F = 0$, the system we will consider throughout the chapter. This line of reasoning, to explain external forces by other dynamical systems, can be investigated in a much deeper way, as done by HERTZ, who developed a theory of mechanics which does not rely on the definition of forces [18]. If only linear equations of motion are considered the above reasoning is however a little bit artificial since limited sources of energy normally yield nonlinear equations of motion. Nevertheless the above setting can simplify the mathematical analysis of problems significantly.

2.2 Underlying Equations of Motion

We consider the differential equations of a conservative system, perturbed by linear parametric periodic excitation, which, without loss of generality, can be written as

$$M\ddot{q} + \Delta D(t, \varepsilon)\dot{q} + (K + \Delta K(t, \varepsilon))q = 0, \tag{2.4a}$$

$$\Delta D(t, \varepsilon) = \Delta D(t + 2\pi, \varepsilon), \qquad \Delta K(t, \varepsilon) = \Delta K(t + 2\pi, \varepsilon), \tag{2.4b}$$

$$M = \mathrm{diag}(1, 1, \ldots, 1), \qquad\qquad K = \mathrm{diag}(\omega_1^2, \ldots, \omega_N^2) \tag{2.4c}$$

by introducing a dimensionless time based on the frequency of the parametric excitation. We assume that $\Delta D(t, \varepsilon)$ and $\Delta K(t, \varepsilon)$ depend smoothly on the parameter ε and can be expanded as

$$\Delta D(t, \varepsilon) = \Delta D_1(t)\varepsilon + \Delta D_2(t)\varepsilon^2 + \cdots, \tag{2.5a}$$

$$\Delta K(t, \varepsilon) = \Delta K_1(t)\varepsilon + \Delta K_2(t)\varepsilon^2 + \cdots. \tag{2.5b}$$

The parameter ε can be regarded as a norm of the perturbation. In the following we work with the first order system

$$\dot{x} = Ax \tag{2.6}$$

$$A = \begin{bmatrix} 0 & I \\ -K - \varepsilon \Delta K_1(t) + \cdots & -\varepsilon \Delta D_1(t) + \cdots \end{bmatrix}, \quad x = \begin{bmatrix} q \\ \dot{q} \end{bmatrix}$$

equivalent to (2.4a). It is well known from FLOQUET theory that the stability of the trivial solution depends on the eigenstructure of the monodromy matrix, which we expand in terms of the perturbation parameter ε

$$X(2\pi, \varepsilon) = X(2\pi, 0) + \frac{\partial X(2\pi, 0)}{\partial \varepsilon}\bigg|_{\varepsilon=0} \varepsilon + \cdots . \tag{2.7}$$

From the perturbation theory developed in [67, 69] it is known that the derivative of the monodromy matrix can be calculated as

$$\frac{\partial X(2\pi, \varepsilon)}{\partial \varepsilon}\bigg|_{\varepsilon=0} = X(2\pi, 0)H, \tag{2.8a}$$

with

$$H = \int_0^{2\pi} Y(t, 0)^T \frac{\partial A(t, \varepsilon)}{\partial \varepsilon}\bigg|_{\varepsilon=0} X(t, 0)dt, \tag{2.8b}$$

an expression which depends only on the derivative of the system matrix of the corresponding first-order system with respect to ε, the monodromy matrix of the unperturbed problem X and the monodromy matrix Y of its adjoint system

$$\dot{y} = -A^T y. \tag{2.9}$$

Since in our case the unperturbed problem has constant coefficients and is decoupled, we have very simple analytic expressions for $X(t, 0)$ and $Y(t, 0)$ namely

$$X = \begin{bmatrix}
\cos \omega_1 t & 0 & \cdots & 0 & \frac{1}{\omega_1}\sin \omega_1 t & 0 & \cdots & 0 \\
0 & \cos \omega_2 t & \cdots & 0 & 0 & \frac{1}{\omega_2}\sin \omega_2 t & \cdots & 0 \\
\vdots & & & \vdots & \vdots & & & \vdots \\
0 & 0 & & \cos \omega_N t & 0 & 0 & & \frac{1}{\omega_N}\sin \omega_N t \\
-\omega_1 \sin \omega_1 t & 0 & \cdots & 0 & \cos \omega_1 t & 0 & \cdots & 0 \\
0 & -\omega_2 \sin \omega_2 t & \cdots & 0 & 0 & \cos \omega_2 t & \cdots & 0 \\
\vdots & & & \vdots & \vdots & & & \vdots \\
0 & 0 & & -\omega_N \sin \omega_N t & 0 & 0 & & \cos \omega_N t
\end{bmatrix} \tag{2.10}$$

and the matrix $Y = X^{-1}$ simply reads

$$
Y = \begin{bmatrix}
\cos\omega_1 t & 0 & \cdots & 0 & \omega_1\sin\omega_1 t & 0 & \cdots & 0 \\
0 & \cos\omega_2 t & \cdots & 0 & 0 & \omega_2\sin\omega_2 t\cdots & & 0 \\
\vdots & & & \vdots & \vdots & & & \vdots \\
0 & 0 & & \cos\omega_N t & 0 & 0 & & \omega_N\sin\omega_N t \\
-\frac{1}{\omega_1}\sin\omega_1 t & 0 & \cdots & 0 & \cos\omega_1 t & 0 & \cdots & 0 \\
0 & -\frac{1}{\omega_2}\sin\omega_2 t\cdots & & 0 & 0 & \cos\omega_2 t & \cdots & 0 \\
\vdots & & & \vdots & \vdots & & & \vdots \\
0 & 0 & & -\frac{1}{\omega_N}\sin\omega_N t & 0 & 0 & & \cos\omega_N t
\end{bmatrix} .
$$

$$(2.11)$$

The spectrum of the unperturbed problem can be either simple or semi-simple. Semi-simple eigenvalues can occur due to the symmetry of the structure, i.e. $\omega_i = \omega_j$, or due to internal or combination resonances. According to [89] simple and semi-simple eigenvalues of the monodromy matrix can be expanded in terms of ε as

$$
\rho = \rho_0 + \frac{\partial\rho}{\partial\varepsilon}\Big|_{\varepsilon=0}\varepsilon + \cdots , \tag{2.12}
$$

where the derivative of the FLOQUET multiplier can be calculated analytically [66]. The derivative of the modulus can then be obtained [67, 69] from

$$
\frac{\partial|\rho|}{\partial\varepsilon} = \frac{1}{|\rho_0|}\operatorname{Re}\left(\bar{\rho}_0\frac{\partial\rho}{\partial\varepsilon}\right). \tag{2.13}
$$

The eigenvalues ρ_{0j}, the eigenvectors $\boldsymbol{u}_j$ of the monodromy matrix $X(2\pi, 0)$ of the unperturbed problem and the eigenvectors $\boldsymbol{v}_j$ of its adjoint $Y(2\pi, 0)$ have a particularly simple form given by

$$
\rho_{0j} = \cos 2\pi\omega_j + \mathrm{i}\sin 2\pi\omega_j \tag{2.14}
$$

and

$$
\boldsymbol{u}_j = \frac{1}{\sqrt{2}}(0, \cdots, 0, \frac{\mathrm{i}}{\omega_j}, 0, \cdots, 0, 1, 0, \cdots, 0)^T , \tag{2.15a}
$$

$$
\boldsymbol{v}_j = \frac{1}{\sqrt{2}}(0, \cdots, 0, -\mathrm{i}\omega_j, 0, \cdots, 0, 1, 0, \cdots, 0)^T , \tag{2.15b}
$$

where the nonzero entries appear in the j-th and $N+j$-th position. The eigenvectors are normalized such that $\boldsymbol{v}_j^T\boldsymbol{u}_j = 1$.

2.3 Perturbation of Simple Eigenvalues

For a simple FLOQUET multiplier ρ_j, the derivative $\frac{\partial \rho_j}{\partial \varepsilon}\Big|_{\varepsilon=0}$, in the following for simplicity denoted by $\frac{\partial \rho_j}{\partial \varepsilon}$, is given by [68]

$$\frac{\partial \rho_j}{\partial \varepsilon} = \frac{v_j^T \frac{\partial X(2\pi,0)}{\partial \varepsilon}\Big|_{\varepsilon=0} u_j}{v_j^T u_j} = \frac{\rho_{0j} v_j^T H u_j}{v_j^T u_j}. \tag{2.16}$$

Due to the simple structure of X, Y, u_j and v_j and using

$$v_j^T \int_0^{2\pi} Y(t,0)^T \frac{\partial A(t,\varepsilon)}{\partial \varepsilon}\Big|_{\varepsilon=0} X(t,0) dt\, u_k$$

$$= \int_0^{2\pi} v_j^T Y(t,0)^T \frac{\partial A(t,\varepsilon)}{\partial \varepsilon}\Big|_{\varepsilon=0} X(t,0) u_k dt, \tag{2.17}$$

the derivative of ρ_j with respect to ε can be calculated as

$$\frac{\partial \rho_j}{\partial \varepsilon} = \frac{1}{2}\left(-\frac{i}{\omega_j}\int_0^{2\pi}\Delta k_{jj}dt - \int_0^{2\pi}\Delta d_{jj}dt\right)\rho_{0j}, \tag{2.18}$$

where Δk_{jj}, Δd_{jj} are the matrix entries on the diagonal of $\Delta K_1(t)$, $\Delta D_1(t)$. It is interesting to note that on the main diagonal of the matrix $\int_0^{2\pi} V^T Y(t,0)^T$ $\frac{\partial A(t,\varepsilon)}{\partial \varepsilon}\Big|_{\varepsilon=0} X(t,0) U dt$, where U (V) is build up form the eigenvectors of the (adjoint) unperturbed problem, the entries are proportional to the mean value of the perturbation matrices. Since

$$\frac{\partial |\rho_j|}{\partial \varepsilon} = \frac{1}{|\rho_{0j}|}\text{Re}\left(\bar{\rho}_{0j}\frac{\partial \rho_j}{\partial \varepsilon}\right) = -\frac{1}{2}\int_0^{2\pi}\Delta d_{jj}dt, \tag{2.19}$$

we see that in the first approximation for simple eigenvalues the modulus of the eigenvalues is not influenced by restoring terms Δk_{jj} and is decreased by dissipative forces Δd_{jj}.

2.4 Perturbation of Semi-Simple Eigenvalues

For two semi-simple FLOQUET multipliers ρ_j and ρ_k, the first term in (2.12) is calculated according to [67, 69, 72] from

$$\det\begin{bmatrix} \rho_{0j} v_j^T H u_j - \frac{\partial \rho_j}{\partial \varepsilon} & \rho_{0j} v_j^T H u_k \\ \rho_{0k} v_k^T H u_j & \rho_{0k} v_k^T H u_k - \frac{\partial \rho_j}{\partial \varepsilon} \end{bmatrix} = 0, \tag{2.20}$$

using again the adjoint problem. Semi-simple eigenvalues can arise due to multi-ple eigenfrequencies, combination resonances or internal resonances. Let us first consider the case of two equal eigenfrequencies of the unperturbed problem.

For a system with $\omega_i = \omega_j = \omega$ due to (2.20) only the matrix entries k, l with $k, l \in \{i, j\}$ enter the first derivative of the FLOQUET multiplier. The derivative of a semi-simple FLOQUET multiplier can therefore be calculated from an equivalent two by two system of the type (2.4a).

We now consider the case of such a two by two system with three parameters κ, δ, γ where

$$\Delta D(t, \varepsilon) = \delta(\varepsilon)D(t) + G(t), \qquad \Delta K(t, \varepsilon) = \kappa(\varepsilon)K(t) + \gamma(\varepsilon)N(t), \qquad (2.21)$$

and $D(t)$, $K(t)$ are symmetric, periodic, positive semi-definite and $G(t)$, $N(t)$ are skew-symmetric and periodic. In our applications the matrix $G(t)$ originates from small gyroscopic terms, which in some rotating systems arise through CORIOLIS effects when the equations of motion are set up with respect to a rotating frame. Therefore the matrix $G(t)$ will most often be constant, and for moderate angular velocities $\Omega \ll \omega_1$ is small, so that it can be regarded as perturbation. According to the assumptions made in the context (2.4a), the parameters can be expanded in terms of ε as

$$\kappa(\varepsilon) = \kappa_1\varepsilon + \kappa_2\varepsilon^2 + \cdots \qquad (2.22a)$$

$$\delta(\varepsilon) = \delta_1\varepsilon + \delta_2\varepsilon^2 + \cdots \qquad (2.22b)$$

$$\gamma(\varepsilon) = \gamma_1\varepsilon + \gamma_2\varepsilon^2 + \cdots . \qquad (2.22c)$$

The derivative of the FLOQUET multiplier can be calculated from

$$\det\left(\int_0^{2\pi} \frac{-\rho_0}{2}\left(\delta D(t) + G(t) + \frac{i}{\omega}\left(\kappa K(t) + \gamma N(t)\right)\right)dt - \frac{\partial\rho}{\partial\varepsilon}I\right) = 0. \qquad (2.23)$$

Assuming that $\int_0^{2\pi} D(t)dt$ is positive definite, which is most often the case due to material damping, the number of parameters in (2.23) can be reduced by performing an orthogonal transformation

$$\det\left(Q^T\left(\int_0^{2\pi} \frac{-\rho_0}{2}\left(\delta D(t) + G(t) + \frac{i}{\omega}\left(\kappa K(t) + \gamma N(t)\right)\right)dt\right)Q - \frac{\partial\rho}{\partial\varepsilon}I\right)$$

$$= \det\left(\frac{-\rho_0}{2}\left(\begin{bmatrix} \delta d & g \\ -g & \delta(d + \Delta d) \end{bmatrix} + \frac{i}{\omega}\begin{bmatrix} \kappa k & \gamma n \\ -\gamma n & \kappa(k + \Delta k) \end{bmatrix}\right) - \frac{\partial\rho}{\partial\varepsilon}\begin{bmatrix} 1 & 0 \\ 0 & 1 \end{bmatrix}\right) \qquad (2.24a)$$

where Q consists of the eigenvectors, which simultaneously diagonalize $\int_0^{2\pi} D(t)dt$ and $\int_0^{2\pi} K(t)dt$. The derivative of the FLOQUET multiplier can thus be calculated as

$$\frac{\partial \rho}{\partial \varepsilon}\Big|_{\varepsilon=0} = \rho_0 \left(-\frac{d\delta}{2} - \frac{\delta \Delta d}{4} - i\left(\frac{k\kappa}{2\omega} + \frac{\Delta k\kappa}{4\omega} \right) \right.$$
$$\left. \pm \sqrt{-\frac{g^2}{4} + \frac{n^2\gamma^2}{4\omega^2} + \frac{\delta^2 \Delta d^2}{16} - \frac{\Delta k^2 \kappa^2}{16\omega^2} + i\left(\frac{\delta \Delta d \Delta k \kappa}{8\omega} - \frac{gn\gamma}{2\omega} \right)} \right)$$

$$(2.25)$$

The first approximation to the stability boundary described by $\frac{\partial |\rho|}{\partial \varepsilon} = 0$ with equation (2.13) reads

$$0 = -\delta\left(\frac{d}{2} + \frac{\Delta d}{4} \right) + \mathrm{Re}\left(\sqrt{-\frac{g^2}{4} + \frac{n^2\gamma^2}{4\omega^2} + \frac{\delta^2 \Delta d^2}{16} - \frac{\Delta k^2 \kappa^2}{16\omega^2} + i\left(\frac{\delta \Delta d \Delta k \kappa}{8\omega} - \frac{gn\gamma}{2\omega} \right)} \right)$$

$$(2.26a)$$

$$= -\delta\left(\frac{d}{2} + \frac{\Delta d}{4} \right) + |z| \cos\left(\frac{1}{2} \arg(z) \right)$$

$$(2.26b)$$

with

$$z = -\frac{g^2}{4} + \frac{n^2\gamma^2}{4\omega^2} + \frac{\delta^2 \Delta d^2}{16} - \frac{\Delta k^2 \kappa^2}{16\omega^2} + i\left(\frac{\delta \Delta d \Delta k \kappa}{8\omega} - \frac{gn\gamma}{2\omega} \right), \quad (2.26c)$$

$$\arg(z) = \arctan\left(\frac{\mathrm{Im}(z)}{\mathrm{Re}(z)} \right). \quad (2.26d)$$

The corresponding graph is shown in Fig. 2.1 for $\omega = 0.6$, $d = \pi$, $n = \pi$, $k = \pi$, $\Delta k = \pi$, $\Delta d = 0.3\pi$. At first we consider the case $g = 0$, which means that no gyroscopic perturbations are present. Assuming additionally $\Delta d = 0$, the expression for the stability boundary simplifies substantially and can be written as

$$\gamma n = \sqrt{\frac{\kappa^2 \Delta k^2}{4} + \delta^2 d^2 \omega^2}. \quad (2.27)$$

The corresponding stability boundary is given in Fig. 2.2 and coincides with the numerical example

$$\begin{bmatrix} 1 & 0 \\ 0 & 1 \end{bmatrix} \ddot{q} + \delta(\varepsilon) \cos^2 t \begin{bmatrix} 1 & 0 \\ 0 & 1 \end{bmatrix} \dot{q} + \left(\begin{bmatrix} \omega^2 & 0 \\ 0 & \omega^2 + \kappa(\varepsilon) \end{bmatrix} + \gamma(\varepsilon) \cos^2 t \begin{bmatrix} 0 & -1 \\ 1 & 0 \end{bmatrix} \right) q = 0$$

$$(2.28)$$

studied in [72] (i.e. $k = 0$, $\Delta k = 2\pi$, $d = \pi$, $\Delta d = 0$, $n = \pi$) in a three parameter setting.

By comparison of Figs. 2.1 and 2.2 it is seen that they are qualitatively equivalent. In Fig. 2.3 cuts through the stability domain at $\kappa_1 = 0$ and $\kappa_1 = 0.13$ ($\frac{\omega_2}{\omega_1} \approx 1.15$)

Fig. 2.1 Approximation
to the stability domain for
$\Delta k = \pi, \Delta d = 0.3\pi$

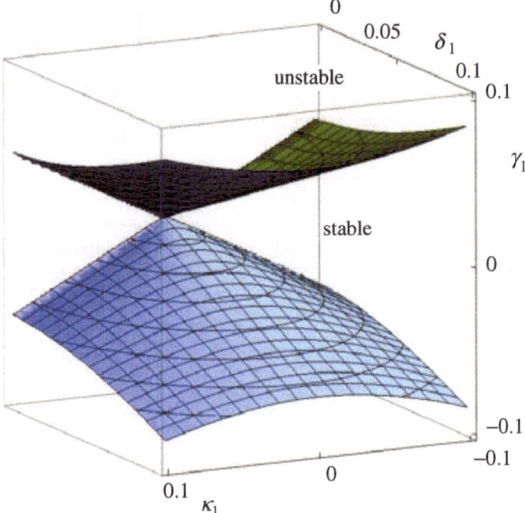

Fig. 2.2 Approximation to
the stability boundary for
$\Delta d = 0$

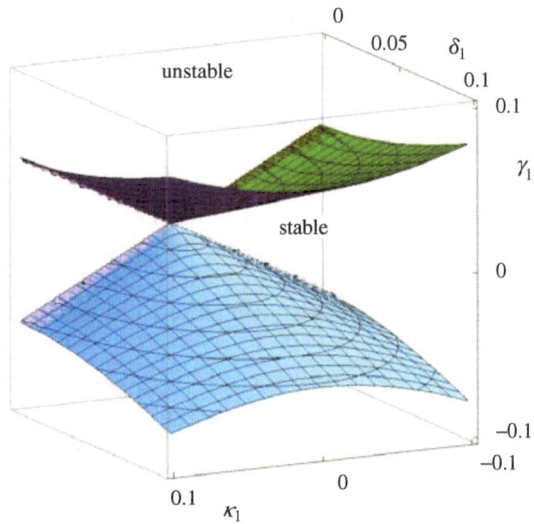

are shown in comparison with a numerical evaluation of the FLOQUET multipliers.
Figure 2.4 shows cuts through the stability region at $\delta_1 = 0$ and $\delta_1 = 0.3$.

It is seen, that for small perturbations the approximation of the stability boundary
is in very good agreement with the numerical results, especially in the technically
relevant range with small damping and a moderate splitting of the frequencies up to
10 percent.

Now we consider the case of $g \neq 0$, i.e. the occurrence of small gyroscopic
terms. Considering the same parameter set as before, $d = \pi, n = \pi, k = \pi, \Delta k = \pi$,

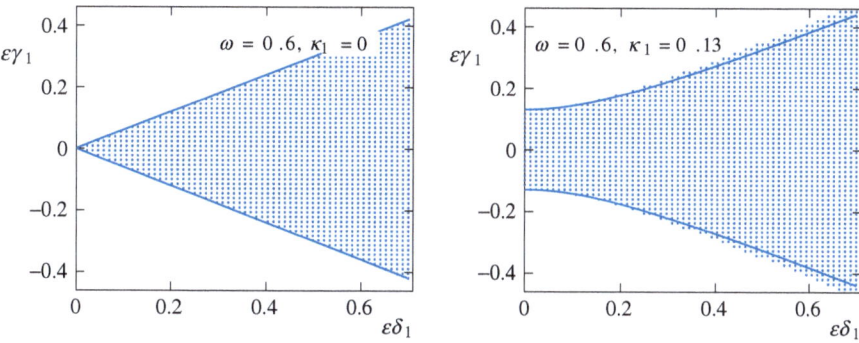

Fig. 2.3   Stable regions for a symmetric and an asymmetric system (*dot* stable)

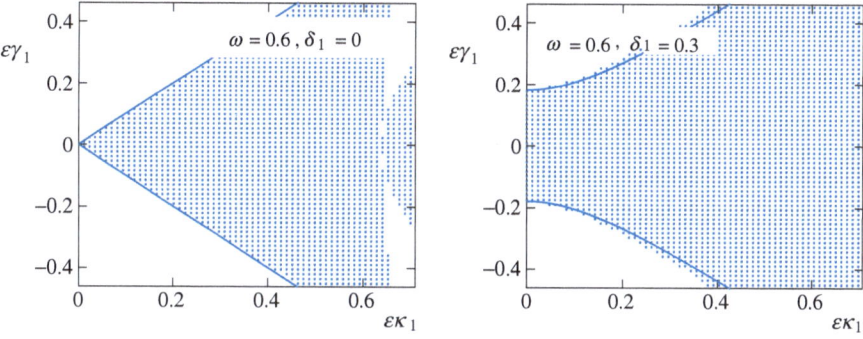

Fig. 2.4   Stable regions for a symmetric and an asymmetric system (*dot* stable)

$\Delta d = 0.3\pi$, with $g = 0.01$ we obtain the approximation of the stability boundary shown in Fig. 2.5. Taking a closer look at the cuts for different values of κ we obtain the curves shown in Fig. 2.6. It is seen that the gyroscopic terms destroy the stability boundary from Fig. 2.4, i.e. the splitting of eigenfrequencies without introduction of damping does not have an effect. However we see that in combination with slight damping also for this case an enlargement of the stable region is achieved. For larger values of δ and γ, the role of the gyroscopic terms is almost negligible since the stability boundary calculated for $\kappa = 0.1$ gets closer to the red curve obtained for $g = 0$ and the same value of κ. For large gyroscopic terms, the effect of stabilization by splitting of eigenfrequencies is destroyed, as we see from Fig. 2.7 where the entries of the gyroscopic matrix are of the order of magnitudes of those of the stiffness matrix. If we have combination resonances i.e. $\omega_i = m\,\omega_j$ for m integer, the analysis can also be performed with (2.20), however the expressions are much lengthier than in the case $m = 1$. From numerical investigations it is seen that the system is less sensitive to self-excited vibrations at combination resonances as in the case $\omega_1 = m\,\omega_2$. This was to be expected, since the stiffness terms in the perturbation expressions are divided by the frequency. The corresponding graph is shown in Fig. 2.8 for $\omega_1 = 0.6$, $\omega_2 = 2\omega_1$,

Fig. 2.5 Approximation to
the stability boundary for
$g \neq 0$

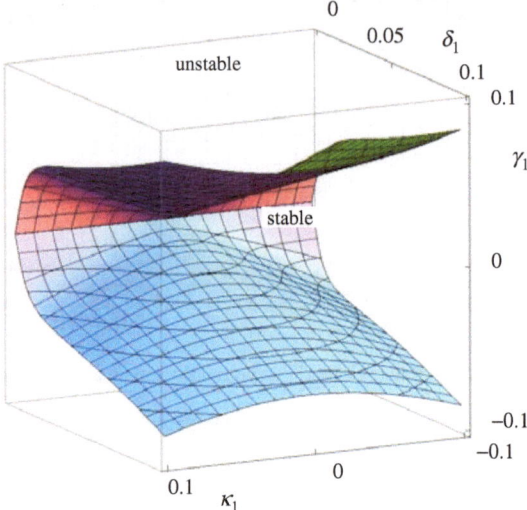

Fig. 2.6 Approximation of
the stability boundary for
$g = 0$ (*red*) and $g = 0.01$ for
$\kappa = 0, 0.08, 0.1$

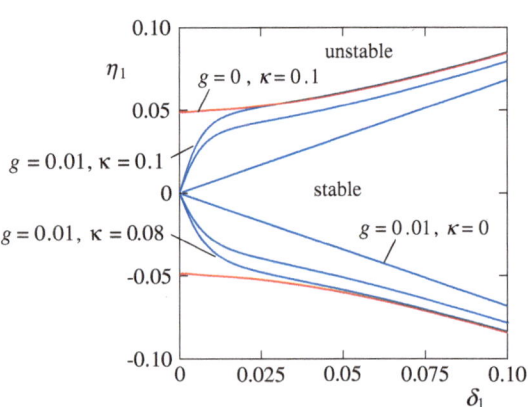

$d = \pi, n = \pi, k = \pi, \Delta k = \pi, \Delta d = 0.3\pi$, in which the unstable region is much narrower than in the corresponding case with double eigenfrequency at $\omega = 0.6$, which is shown in Fig. 2.1.

Let us now consider the case of an internal resonance for the jth eigenvalue, i.e. we have $\omega_j = \frac{k}{2}$ such that the imaginary part of the FLOQUET multiplier in (2.14) vanishes. Instead of a pair of complex conjugate FLOQUET multipliers we now have a double eigenvalue $\rho_0 = (-1)^k$ which is semi-simple. The first derivative of the FLOQUET multipliers can again be analyzed from (2.20). Examining the corresponding eigenvectors, we observe that the formulas for the first derivative of the FLOQUET multipliers coincide with the ones for a damped version of HILL's equation

$$\ddot{q} + \varepsilon \Delta d_{jj}(t)\dot{q} + (\omega_j^2 + \varepsilon \Delta k_{jj}(t))q = 0, \tag{2.29}$$

Fig. 2.7 Approximation to the stability boundary for $g = \pi$

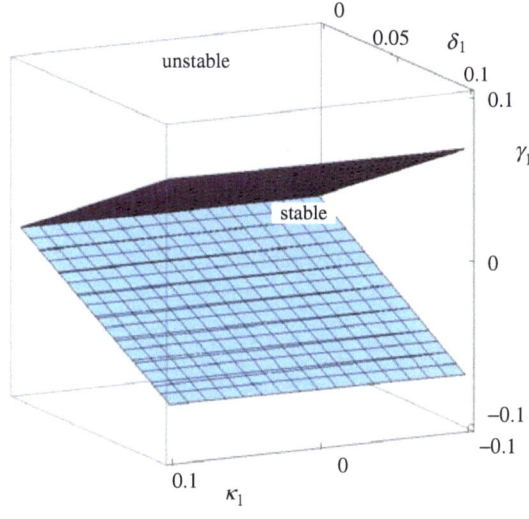

which have been calculated in [65] for a constant damping term. For a periodic damping term, the first derivatives of the corresponding multipliers read

$$\frac{\partial \rho}{\partial \varepsilon} = (-1)^k \left(-c_d k \pm \pi \sqrt{(2a_k - b_d k)^2 + (2b_k - a_d k)^2 - (2c_k)^2} \right), \qquad (2.30)$$

where

Fig. 2.8 Approximation to the stability domain for $\Delta k = \pi,\ \Delta d = 0.3\pi$

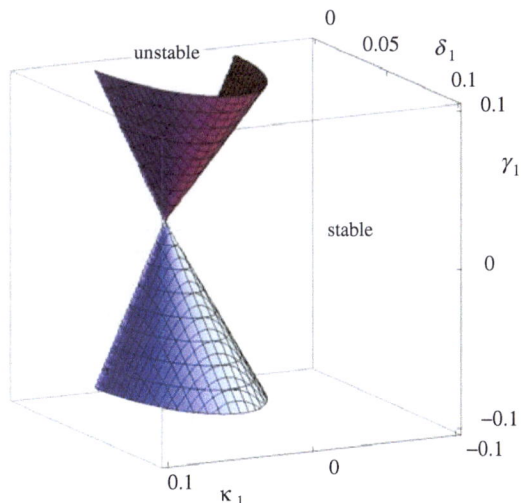

$$a_k = \frac{1}{2\pi k} \int_0^{2\pi} \sin(k\tau)\Delta k_{jj}(\tau)\mathrm{d}\tau, \qquad a_d = \frac{1}{2\pi k} \int_0^{2\pi} \sin(k\tau)\Delta d_{jj}(\tau)\mathrm{d}\tau,$$

$$b_k = \frac{1}{2\pi k} \int_0^{2\pi} \cos(k\tau)\Delta k_{jj}(\tau)\mathrm{d}\tau, \qquad b_d = \frac{1}{2\pi k} \int_0^{2\pi} \cos(k\tau)\Delta d_{jj}(\tau)\mathrm{d}\tau,$$

$$c_k = \frac{1}{2\pi k} \int_0^{2\pi} \Delta k_{jj}(\tau)\mathrm{d}\tau, \qquad c_d = \frac{1}{2\pi k} \int_0^{2\pi} \Delta d_{jj}(\tau)\mathrm{d}\tau.$$

From the truncated expansion

$$\rho_j = \rho_{0j} + \frac{\partial \rho_j}{\partial \varepsilon} + \mathcal{O}(\varepsilon) = (-1)^k \left(1 + \left(-c_d k \pm \pi\sqrt{D}\right)\right) + \mathcal{O}(\varepsilon), \qquad (2.31)$$
$$D = (2a_k - b_d k)^2 + (2b_k - a_d k)^2 - (2c_k)^2,$$

of the FLOQUET multiplier we observe that the system tends to get unstable if the term under the square root is larger than $|c_d k|$. In particular, this is the case when damping is absent and describes vertices of the instability regions for the HILL and the MATHIEU equation [65, 68]. From (2.30) we see that damping has a stabilizing effect, provided

$$c_d > \sqrt{a_d^2 + b_d^2}.$$

For increasing k the damping terms dominate, since they are multiplied by k.

2.5 Discussion

The results obtained in the previous sections obviously only have a local character. They show that, in the vicinity of a semi-simple eigenvalue, a skew symmetric perturbation has a particularly strong destabilizing effect. At the same time it has to be noted that for $\frac{\partial |\rho|}{\partial \varepsilon} = 0$, ρ moves along the unit circle in the complex plane in the first approximation. The system can still be destabilized by higher order terms. In particular this is the case for constant coefficient M, ΔG, K, ΔN systems with very small gyroscopic terms that are almost always unstable [24, 33, 40] although neither ΔG nor ΔN appear in the first derivative of the expansion for the FLOQUET multipliers, provided the spectrum of the unperturbed problem is simple.

Only in specific cases, for example for systems with reversible symmetry (i.e. invariant under time reversal), we can conclude that in the absence of damping the system has a double eigenvalue at the stability boundary [96]. In [96] p. 133 it is shown that if $\Delta D(-t) = -\Delta D(t)$ and $\Delta K(-t) = \Delta K(t)$, then the system is invariant under time reversal.

Despite of having only local character, the perturbation formulas indicate that splitting up the eigenfrequencies of the unperturbed problem tends to stabilize the system, especially when additional damping is present. In order to design a robust system against self-excited vibrations one should therefore maximize the minimal distance between two neighboring eigenfrequencies. This is the goal of structural optimization approaches of later chapters, and the mathematical structure of such an optimization problem is investigated in the next chapter.

Chapter 3
Eigenvalue Placement for Structural Optimization

The purpose of this chapter is to revise relevant literature for eigenvalue problems and to investigate the mathematical properties of the structural optimization problem studied in later chapters.

3.1 Introduction

In the previous section we saw that multiple eigenfrequencies can cause self-excited vibrations for a relatively wide class of systems. For a robust design with respect to self-excitation it is therefore advisable to split eigenfrequencies as much as possible. In this section we will formulate this as an optimization problem and will investigate its mathematical properties. The easiest way to avoid repeated eigenvalues, would be to assign its eigenvalues a priory, which reminds of the pole placement problem in control theory. It is well known that for a controllable system all poles can be placed as desired by an appropriate feedback loop e.g. [86]. However, this procedure requires actuators and is therefore not appropriate for our goals.

For a purely passive system, the placement of eigenvalues is a much more difficult problem. Passive placement of eigenvalues in a mechanical context is often concerned with the avoidance of resonance frequencies [3, 39, 51, 52] and is applied in the context of receptance based methods, which work particularly well if only one frequency is to be shifted.

Other eigenvalue approaches in mechanics are concerned with maximizing the minimum eigenvalue of a mechanical system, the most famous example probably being the search for the optimal column [68]. This problem has received much attention, since the optimal solution occurs at a double eigenvalue which means that the objective function is not differentiable at the optimum.

A very helpful property for many problems in eigenvalue optimization is that the maximum eigenvalue as well as the sum of the maximal eigenvalues are convex functions [4, 61, 62] over the cone of positive semidefinite matrices. This property

G. Spelsberg-Korspeter, *Robust Structural Design against Self-Excited Vibrations*, SpringerBriefs in Applied Sciences and Technology, DOI: 10.1007/978-3-642-36552-2_3, © The Author(s) 2013

makes it possible to formulate the problem of minimizing the sum of the largest eigenvalues as a semidefinite optimization problem, which is basically a generalization of linear optimization to matrix valued functions. A helpful introduction into semidefinite programming and the corresponding duality theory can be found in [17, 88], a Matlab package has been developed in [84]. Unfortunately, however, the nice properties of the sums of eigenvalues do not carry over to their differences, which we have to study in the context of structural optimization. Even in the more recent survey articles on eigenvalue optimization, e.g. [43, 44, 60], this problem has not been mentioned and is therefore investigated in the following.

3.2 Physical Motivation of Eigenvalue Optimization

The linearized equations of motion of a discrete, conservative system read

$$M\ddot{q} + Kq = 0, \tag{3.1}$$

with M and K symmetric, positive definite, where the entries of the stiffness matrix k_{ij} can be interpreted as linear springs that are stretched by the relative motion between two coordinates q_i and q_j. Since actio equals reactio, the contribution to the stiffness matrix of each spring is a positive semi-definite matrix. The eigenvalue problem for the free vibrations can be written as

$$\det(-I\omega^2 + M^{-\frac{1}{2}}KM^{-\frac{1}{2}}) = \det(-\lambda I + A) = 0, \tag{3.2}$$

with $A = M^{-\frac{1}{2}}KM^{-\frac{1}{2}}$, $\lambda = \omega^2$, where $A = \sum_{i=1}^{N} \alpha_i A_i$ is a linear combination of positive semidefinite matrices corresponding to the springs. Therefore, in order to optimize eigenvalues with respect to stiffness modifications, it is sufficient to consider a linear combination of matrices. If modifications both in the mass and stiffness matrix are to be considered, as for example in continuous systems, the matrix A will be a more complicated function of the parameters. In the sequel we will however first investigate the mathematical properties of stiffness modifications only, and afterwards draw conclusions for the general case. Before we consider the optimization of the differences of eigenvalues, we will derive a necessary design condition for the avoidance of multiple eigenfrequencies.

3.3 Relation of Symmetry and Multiple Eigenvalues

In this section we will show that every elastic body with a cyclic symmetry with an angle less than π has at least one double eigenfrequency. A body with such a symmetry can be divided in $n > 2$ equal sectors, which are invariant under rotation about the angle of $2\pi/n$. For the eigenvalue problem of a matrix A a symmetry means that a permutation matrix T exists such that

$$TA = AT, \tag{3.3}$$

i.e. A commutes with T and $A = T^{-1}AT$. For the eigenvalue problem

$$(A - \lambda I)v = 0 \tag{3.4}$$

it follows that

$$(T^{-1}AT - \lambda T^{-1}IT)v = T^{-1}(A - \lambda I)Tv = 0, \tag{3.5}$$

which means that if v is an eigenvector corresponding to λ then also Tv is. It remains to be shown that $v \neq Tv$ at least for some eigenvalues. This can be done by exploiting the eigenvalues and eigenvectors of T. In the following we concentrate on a structure with cyclic symmetry of an angle $2\pi/n$ with $n > 2$. If the degrees of freedom of the structure are defined and grouped accordingly, the matrix T is a block diagonal matrix of cyclic permutation matrices, i.e.

$$T = \begin{bmatrix} T_1 & 0 & \cdots \cdots & & 0 \\ 0 & T_2 & 0 & \cdots & & 0 \\ \vdots & & & & & \vdots \\ 0 & \cdots & 0 & T_{N-1} & 0 \\ 0 & & \cdots \cdots & 0 & T_N \end{bmatrix}, \tag{3.6}$$

where N is the number of different types of generalized coordinates, which are transformed into one another by the symmetry. Each matrix T of dimension n has the form

$$T_i = \begin{bmatrix} 0 & 0 & \cdots & 0 & 1 \\ 1 & 0 & 0 & \cdots & 0 \\ 0 & 1 & 0 & \cdots & 0 \\ \vdots & & & & \vdots \\ 0 & \cdots & 0 & 1 & 0 \end{bmatrix}. \tag{3.7}$$

From the expansion theorem of the determinant it follows that the characteristic equation of T_i reads $\lambda^n - 1 = 0$, therefore (c.f. [82]) the eigenvalues are

$$\lambda_k = e^{ik\frac{2\pi}{n}} \quad k = 0, 1, \ldots, n-1 \tag{3.8}$$

with corresponding eigenvectors

$$v_k = [1, \lambda_k^{n-1}, \ldots, \lambda_k]^T. \tag{3.9}$$

If we have a cyclic symmetry with an angle less than π, i.e. $n \geq 2$, it follows that the vectors $\boldsymbol{T}\boldsymbol{v}$ are not collinear with $\boldsymbol{v}$ and we have proved that we have in fact multiple semi-simple eigenvalues. The knowledge of the structure of $\boldsymbol{T}$ can however be used further. We define $\boldsymbol{Q} = [\boldsymbol{v}_1, \ldots, \boldsymbol{v}_{Nn}]$, i.e. build a matrix of the eigenvectors of $\boldsymbol{T}$ from the augmented eigenvectors of $\boldsymbol{T}_i$. From $\boldsymbol{A} = \boldsymbol{T}^{-1}\boldsymbol{A}\boldsymbol{T}$ it follows

$$\boldsymbol{v}_k^*\boldsymbol{A}\boldsymbol{v}_l = \lambda_k^*\boldsymbol{v}_k^*\boldsymbol{A}\boldsymbol{v}_l\lambda_l, \tag{3.10}$$

which implies that

$$\boldsymbol{v}_k^*\boldsymbol{A}\boldsymbol{v}_l = 0 \quad k \neq l \text{ and } \lambda_k \neq \lambda_l. \tag{3.11}$$

The transformation

$$\tilde{\boldsymbol{A}} = \boldsymbol{Q}^*\boldsymbol{A}\boldsymbol{Q} \tag{3.12}$$

after possible reordering therefore brings $\boldsymbol{A}$ into block diagonal form.

The above reasoning only used arguments form linear algebra for this special type of cyclic symmetry. A much more general picture is given by mathematical group theory [64, 82]. Any group has an irreproducible representation with the knowledge of which a basis can be constructed in which the matrix $\boldsymbol{A}$ decouples into block diagonal form. The size and the multiplicity of the blocks can be calculated using only information of the symmetry group. This procedure is often used to reduce computational effort in the analysis of problems with symmetries.

As an example, finally consider the planar system shown in Fig. 3.1 consisting of 4 equal particles connected by springs with stiffnesses c and k respectively. Each particle i can move in the x and y direction. In order to avoid rigid body modes each particle is restrained for movement with respect to the ground by springs of stiffness

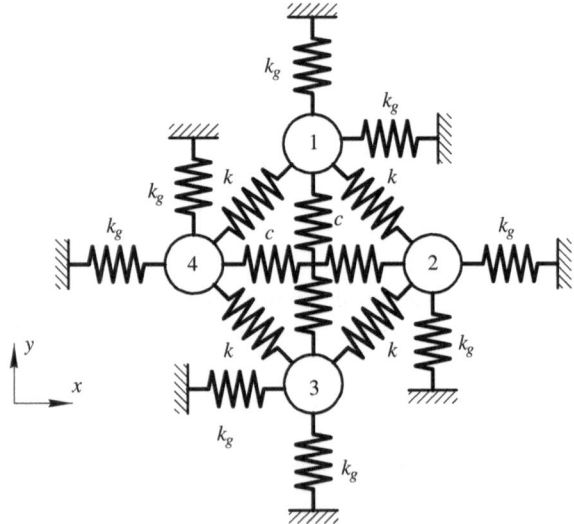

Fig. 3.1 System with double symmetry plane

k_g in x and y direction. If the degrees of freedom are ordered such that

$$\boldsymbol{q} = [q_{1x}, q_{2x}, q_{3x}, q_{4x}, q_{1y}, q_{2y}, q_{3y}, q_{4y}]^T, \tag{3.13}$$

the stiffness matrix of the corresponding equations of motion is

$$\boldsymbol{K} = \begin{bmatrix}
k + k_g & -\frac{k}{2} & 0 & -\frac{k}{2} & 0 & \frac{k}{2} & 0 & -\frac{k}{2} \\
-\frac{k}{2} & c + k + k_g & -\frac{k}{2} & -c & \frac{k}{2} & 0 & -\frac{k}{2} & 0 \\
0 & -\frac{k}{2} & k + k_g & -\frac{k}{2} & 0 & -\frac{k}{2} & 0 & \frac{k}{2} \\
-\frac{k}{2} & -c & -\frac{k}{2} & c + k + k_g & -\frac{k}{2} & 0 & \frac{k}{2} & 0 \\
0 & \frac{k}{2} & 0 & -\frac{k}{2} & c + k + k_g & -\frac{k}{2} & -c & -\frac{k}{2} \\
\frac{k}{2} & 0 & -\frac{k}{2} & 0 & -\frac{k}{2} & k + k_g & -\frac{k}{2} & 0 \\
0 & -\frac{k}{2} & 0 & \frac{k}{2} & -c & -\frac{k}{2} & c + k + k_g & -\frac{k}{2} \\
-\frac{k}{2} & 0 & \frac{k}{2} & 0 & -\frac{k}{2} & 0 & -\frac{k}{2} & k + k_g
\end{bmatrix}. \tag{3.14}$$

The matrix $\boldsymbol{T}$ in this case reads

$$\boldsymbol{T} = \begin{bmatrix}
0 & 0 & 0 & 0 & 0 & 0 & 0 & 1 \\
0 & 0 & 0 & 0 & 1 & 0 & 0 & 0 \\
0 & 0 & 0 & 0 & 0 & 1 & 0 & 0 \\
0 & 0 & 0 & 0 & 0 & 0 & 1 & 0 \\
0 & 0 & 0 & -1 & 0 & 0 & 0 & 0 \\
-1 & 0 & 0 & 0 & 0 & 0 & 0 & 0 \\
0 & -1 & 0 & 0 & 0 & 0 & 0 & 0 \\
0 & 0 & -1 & 0 & 0 & 0 & 0 & 0
\end{bmatrix}, \tag{3.15}$$

which by straight forward redefinition of the degrees of freedom can be brought into the form of (3.6). Using (3.12) the eigenvalues of $\boldsymbol{K}$ can be determined analytically as $\{k_g, k_g, k_g, 2c + k_g, 2k + k_g, 2k + k_g, 2k + k_g, 2c + 2k + k_g\}$, which are not all distinct.

So far we have only considered cyclic symmetries for structures with a finite number of degrees of freedom. However the above reasoning carries over to continuous structures either through discretization, for example with finite elements, or to the calculation of eigenvalues through the characteristic equation. For a symmetric distribution of material and geometrical parameters on a domain with cyclic symmetry the corresponding linearized partial differential equations are invariant with respect to rotation of an angle $2\pi/n$. The eigenvalues are determined form the characteristic equation which is obtained by adjusting the linear combination of the fundamental solutions $\boldsymbol{f}_{ik}(\boldsymbol{x})$ on each domain i

$$\boldsymbol{w}_i = \sum_k C_{ik} \boldsymbol{f}_{ik}(\boldsymbol{x}), \tag{3.16}$$

where the C_{ik} are constants, to the boundary and transition conditions. This yields a transcendental eigenvalue problem of the form

$$A(\lambda)C = 0. \tag{3.17}$$

If the boundary and transition condition also have a cyclic symmetry the same reasoning as for the matrix eigenvalue problem (3.4) can be done, i.e. a permutation matrix T exists such that

$$TA(\lambda) = A(\lambda)T. \tag{3.18}$$

The structure of T is the same as (3.6) and analogous conclusions to the above reasoning can be drawn.

3.4 Mathematical Background on the Optimization of the Differences of Eigenvalues

The goal of this section is to investigate the optimization problem of the differences of eigenvalues from a mathematical point of view, in order to get a better understanding of the underlying topology of the admissible solutions. To this end consider the eigenvalue problem

$$\left(A_0 + \sum_{i=1}^{N} \alpha_i A_i - \lambda I \right) v = 0, \tag{3.19}$$

where the matrices A_i are real, symmetric and positive semidefinite and $0 \leq \alpha_i \leq 1$. We now want to choose the α_i such that the square of the minimal difference between eigenvalues is maximized. This is stated in the optimization problem

$$\max_{\alpha} \min_{ij} (\lambda_i - \lambda_j)^2. \tag{3.20}$$

At first consider matrices of dimension two by two and $N = 1$, the eigenvalue problem then reads

$$\det \left(\begin{bmatrix} a_0 & c_0 \\ c_0 & b_0 \end{bmatrix} + \alpha_1 \begin{bmatrix} a_1 & c_1 \\ c_1 & b_1 \end{bmatrix} - \lambda \begin{bmatrix} 1 & 0 \\ 0 & 1 \end{bmatrix} \right) = 0. \tag{3.21}$$

Without loss of generality we can assume that $c_0 = 0$ and obtain

$$(\lambda_1 - \lambda_2)^2 = [(a_0 - b_0) + \alpha_1(a_1 - b_1)]^2 + 4\alpha_1^2 c_1^2, \tag{3.22}$$

which, by introducing Δ_0 and Δ_1 as the differences of the eigenvalues of the matrices A_0 and A_1 respectively, can be written as

$$\Delta^2 = (\lambda_1 - \lambda_2)^2 = \Delta_0^2 + \alpha_1^2 \Delta_1^2 + 2\alpha_1(a_0 - b_0)(a_1 - b_1), \tag{3.23}$$

which is a parabola of α_1. Since $\Delta_1^2 > 0$, the maximum is attained on the boundary either at $\alpha_1 = 0$ or $\alpha_1 = 1$. If we consider the optimization problem with matrices $A_1, \ldots, A_N$ sequentially, it is clear that the for all i, α_i is binary and we have combinatorialized the optimization problem.

In the general case of n times n matrices the problem gets more involved since the dependence of Δ on the α_i is no longer quadratic and not convex. However we will be able to get at least some insight into the topology of the corresponding optimization problem

$$\max_{\alpha} \min_{i}(\lambda_{i+1} - \lambda_i), \qquad \lambda_1 \leq \lambda_2 \ldots \leq \lambda_N. \tag{3.24}$$

It is well known that the sum of the largest eigenvalues of a symmetric matrix is convex over the cone of n dimensional positive semidefinite symmetric matrices S_n [20, 61]. Therefore, the largest eigenvalues of the matrix

$$A = A_0 + \sum_{i=1}^{N} \alpha_i A_i, \qquad (\alpha_1, \ldots, \alpha_N) \in \mathbb{R}^N \tag{3.25}$$

form a convex function over $\mathbb{R}^N$, namely

$$\sigma_p = \sum_{i=p}^{N} \lambda_i(A) \tag{3.26}$$

is a convex function over $\mathbb{R}^N$, where $\lambda_i(A) = \lambda_i(\alpha_1, \ldots, \alpha_N)$. From the definition of σ_p it follows that

$$\sigma_1 = \text{Tr}(A) \tag{3.27}$$

is a linear function over $\mathbb{R}^N$. The function of the difference of neighboring eigenvalues is given by

$$\Delta_p = \lambda_{p+1} - \lambda_p = 2\sigma_p - \sigma_{p-1} - \sigma_{p+1}, \tag{3.28}$$

which is the difference of two convex functions, namely $\sigma_{p-1} + \sigma_{p+1}$ and $2\sigma_p$. If we limit our investigation to the case $n = 2$, we obtain

$$\Delta_1 = 2\sigma_2 - \sigma_1 = \lambda_2 - \text{Tr}(A). \tag{3.29}$$

Since $\text{Tr}(A)$ is a linear function in α and therefore has no curvature, and the maximum eigenvalue λ_2 is a convex function, it is clear that also Δ_1 is a convex function in $\mathbb{R}^N$ which assumes its maxima on the boundaries for α_i.

Unfortunately the general case of $n > 2$ is much more complicated. Even though Δ_p is the difference of two convex functions this does not mean that it is convex. In the literature, optimization problems for differences of two convex functions are termed d.c. optimization problems, and have been studied in the context of non-convex optimization (see e.g.[14, 15, 45, 83]). Using methods from convex analysis, one can derive optimality criteria and algorithms for global d.c. optimization [14, 83]. However these methods are involved and not directly applicable to our problem, since we have to solve the problem

$$\max_{\alpha} \min_{p} \Delta_p. \tag{3.30}$$

In the max-min problem not only the local maxima of Δ_p are candidates for the optimal solution but also the intersections of the difference curves. This can be easily seen by considering the case of symmetric, positive definite random matrices A_0 and A_1 with dimension $n = 6$ and $N = 1$. Without loss of generality, we can assume A_0 to be a diagonal matrix. It is seen from Fig. 3.2, that although $\sigma_p(\alpha)$ is convex, the differences $\Delta_p(\alpha)$ contain local optima and intersect each other. In this particular example the optimal value of $\Delta_p(\alpha)$ is attained approximately at $\alpha_1^* = 0.18$. Furthermore it is seen that the numerical derivative $\sigma_p'(\alpha)$ of $\sigma_p(\alpha)$ is clearly not convex, which means that also the number of local optima of $\Delta_p(\alpha)$ cannot be bounded a priory. Since $\lambda_j(\alpha)$ is differentiable everywhere except for points where the eigenvalues cross, the functions $\sigma_p(\alpha)$ and $\Delta_p(\alpha)$ are piecewise differentiable. For this particular example, the optimum of (3.30) is clearly at an intersection point of

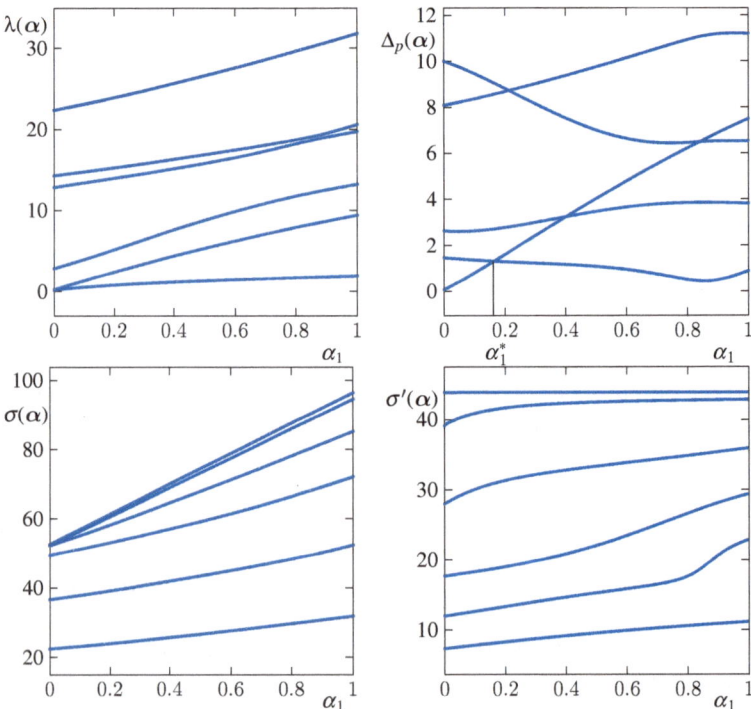

Fig. 3.2 Random example for (3.30) with $n = 6$, $N = 1$

$\Delta_p(\boldsymbol{\alpha})$. The graphs of the functions $\Delta_p(\boldsymbol{\alpha})$ are $N-1$ dimensional manifolds in $\mathbb{R}^N$. Their intersections are $N-2$ dimensional manifolds, which are hard to calculate even for moderate dimensional problems. Summarizing, the above investigation reveals the complex topology underlying the optimization problem (3.30) and shows the difficulties in the development of an algorithm, which is guaranteed to converge to the global optimum. Since the suppression of self-excited vibrations in Chap. 2 was only motivated locally, we do not further pursue the development of a globally convergent algorithm but use heuristics and appropriate local minima.

Chapter 4
Passive Stabilization of Discrete Systems

In this chapter we discuss the possibility to design discrete structures in a robust manner against self-excited vibrations. In order to make use of the results derived in Chap. 2, we have to bring the equations of motion into the form

$$M\ddot{q} + \Delta D(t, \varepsilon)\dot{q} + (K + \Delta K(t, \varepsilon))q = 0. \tag{4.1}$$

In order to achieve this, we have to define appropriate generalized coordinates. Since we are interested in stability investigations, it is worthwhile to investigate the possible influence of the choice of coordinates on the stability behavior.

4.1 Dependence of the Stability on the Choice of the Generalized Coordinates

In the following we investigate an example of a system, in which a motion is stable with respect to one particular set of generalized coordinates and unstable with respect to another. Consider the well known system of a mass on a conveyor belt shown in Fig. 4.1. Between the mass and the belt we assume COULOMB friction with a falling friction characteristic. In the generalized coordinate q the equations of motion linearized around the equilibrium position read

$$m\ddot{q} + d\dot{q} + kq = 0, \tag{4.2}$$

where $d < 0$, which originates from the falling friction characteristic, destabilizes the trivial solution, as seen from an eigenvalue analysis. However if we observe the system from a moving observer measuring the angle φ, with

$$\tan \varphi = \frac{q}{x}, \tag{4.3}$$

G. Spelsberg-Korspeter, *Robust Structural Design against Self-Excited Vibrations*, SpringerBriefs in Applied Sciences and Technology, DOI: 10.1007/978-3-642-36552-2_4, © The Author(s) 2013

Fig. 4.1 Mass on conveyor
belt

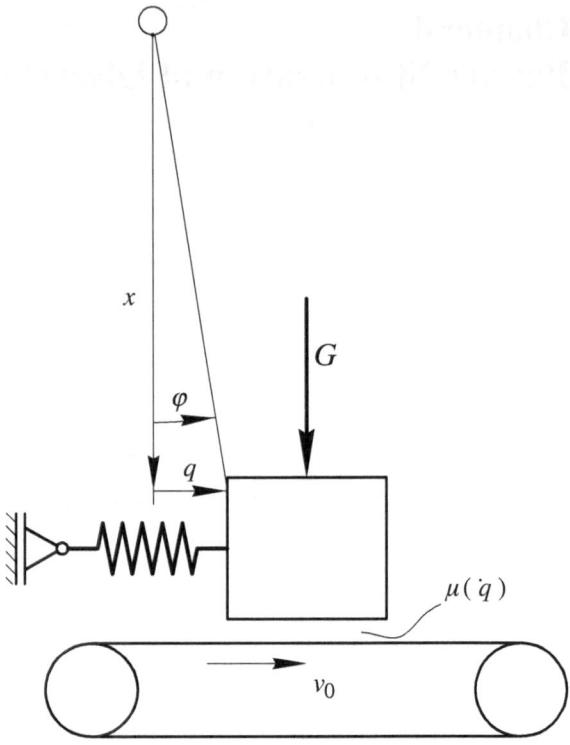

we obtain the linearized equations of motion

$$m(\ddot{x} + 2\dot{x}\dot{\varphi} + \ddot{\varphi}) + d(x\dot{\varphi} + \dot{x}\varphi) + kx\varphi = 0. \tag{4.4}$$

If we specify the position of the observer to be $x = x_0 e^{at}$ and perform an eigenvalue
analysis with the ansatz $\varphi = v e^{\lambda t}$, we obtain

$$\lambda = -\frac{d}{2m} - a \pm \sqrt{\frac{d^2}{4m^2} - k} \tag{4.5}$$

from which we see that even for negative d all eigenvalues have a negative real part if
$a > \frac{|d|}{2m}$. The trivial solution of the system is stable in φ but unstable in q. Note that by
choosing a negative and d positive also the converse is possible. The example shows
that at least theoretically stability depends on the choice of generalized coordinates.
In order to circumvent this problem it is appropriate to reformulate the stability
definition of LIAPOUNOV in terms of a vector norm in the embedding space for the
mechanical system instead of taking the norm of the generalized coordinates. In
contrast to the generalized coordinates which are often chosen as the components

of a vector with respect to a certain basis, direction and length vectors of bodies are invariant.

In order to perform a meaningful stability analysis in the norm of the generalized coordinates one has to assure that the norm is equivalent to the norm of the corresponding vector space in which the physical problem is formulated. The above example shows that a reference to the equivalence of norms in finite dimensional spaces is not sufficient, since the transformations between generalized coordinates are time dependent. For all the systems studied in the sequel, the mentioned equivalence of norms is obvious, therefore it is not shown explicitly.

4.2 Minimal Model for Self-Excited Vibrations of a Rotor in Frictional Contact

In this section we consider the example of a minimal model for a discrete rotor model in frictional contact, depicted in Fig. 4.2. The model can be interpreted as a minimal model for a drum brake or a centrifugal clutch. We consider a rotor of mass M rotating at constant angular velocity Ω. The bearings are modeled as elastic springs with stiffness c_x, c_y respectively. The rotor is in contact with a circular guidance through massless pins which are pressed onto it by linear elastic springs with spring constant k and prestressed with the force N_0. Between the ring and the pins, friction occurs, which we describe using COULOMB's law with constant μ. We assume that the

Fig. 4.2 Minimal model for a rotor in frictional contact

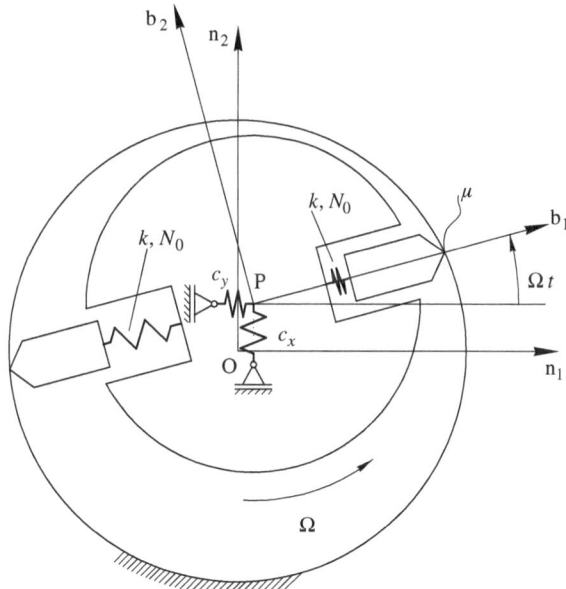

pins are always in contact with the outer ring and that there is always a nonvanishing relative velocity between the ring and the pins. There are two straight forward ways to define the two coordinates of the system. The first one is to express the displacement in the frame B attached to the rotor, i.e. to define the displacement vector as

$$r = \overrightarrow{OP} = \bar{q}_1 b_1 + \bar{q}_2 b_2. \tag{4.6}$$

The linearized equations of motion are then given by

$$
\begin{bmatrix} M & 0 \\ 0 & M \end{bmatrix} \begin{bmatrix} \ddot{\bar{q}}_1 \\ \ddot{\bar{q}}_2 \end{bmatrix} + \begin{bmatrix} 0 & -2M\Omega \\ 2M\Omega & 0 \end{bmatrix} \begin{bmatrix} \dot{\bar{q}}_1 \\ \dot{\bar{q}}_2 \end{bmatrix} +
$$
$$
\left(\begin{bmatrix} k_{11}(t) & k_{12}(t) \\ k_{21}(t) & k_{22}(t) \end{bmatrix} + \begin{bmatrix} \Delta k_{11} & \Delta k_{12} \\ \Delta k_{21} & \Delta k_{22} \end{bmatrix} \right) \begin{bmatrix} \bar{q}_1 \\ \bar{q}_2 \end{bmatrix} = \mathbf{0}, \tag{4.7}
$$

with

$$k_{11}(t) = c_x - M\Omega^2 + (c_y - c_x)\sin^2 \Omega t, \quad k_{12}(t) = (c_y - c_x)\cos \Omega t \sin \Omega t,$$
$$k_{21}(t) = (c_y - c_x)\cos \Omega t \sin \Omega t, \quad\quad\quad k_{22}(t) = c_y - M\Omega^2 + (c_x - c_y)\sin^2 \Omega t$$

and

$$\Delta k_{11} = 2k, \quad \Delta k_{12} = 0,$$
$$\Delta k_{21} = 2k\mu, \quad \Delta k_{22} = 2\frac{N_0}{r}(1 + \mu^2).$$

We observe that the equations of motion have a particularly simple form if the bearing stiffnesses c_x and c_y are equal. Then we obtain

$$k_{11} = c + 2k - M\Omega^2, \quad k_{12} = 0, \quad k_{21} = 2k\mu, \quad k_{22} = c - M\Omega^2 + 2\frac{N_0}{r}(1 + \mu^2)$$

such that the equations of motion have constant coefficients and can be written in the form

$$M\ddot{q} + G\dot{q} + (K + N)q = 0, \tag{4.8}$$
$$M = M^T > 0, \quad G = -G^T, \quad K = K^T, \quad N = -N^T.$$

Since we are dealing with a two by two system, it can be shown analytically that the trivial solution of (4.8) is unstable [24, 33, 40]. The resulting self-excited vibrations can be interpreted as the reason for squeal.

In case of asymmetric bearing stiffness it is no longer obvious weather the trivial solution of (4.8) is stable, since the equations of motion have periodic coefficients. In this case it is beneficial to define the coordinates of the system with respect to the inertial frame N i.e. to define the displacement vector as

$$r = \overrightarrow{OP} = q_1 n_1 + q_2 n_2. \tag{4.9}$$

The equations of motion then read

$$\begin{bmatrix} M & 0 \\ 0 & M \end{bmatrix} \begin{bmatrix} \ddot{q}_1 \\ \ddot{q}_2 \end{bmatrix} + \left(\begin{bmatrix} c_x & 0 \\ 0 & c_y \end{bmatrix} + \begin{bmatrix} \Delta k_{11}(t) & \Delta k_{12}(t) \\ \Delta k_{21}(t) & \Delta k_{22}(t) \end{bmatrix} \right) \begin{bmatrix} q_1 \\ q_2 \end{bmatrix} = \mathbf{0}, \tag{4.10}$$

with

$$\Delta k_{11}(t) = 2k \cos^2 \Omega t + 2 \frac{N_0}{r}(1 + \mu^2) \sin^2 \Omega t - k\mu \sin 2\Omega t,$$

$$\Delta k_{12}(t) = 2k \cos \Omega t \sin \Omega t - 2 \frac{N_0}{r}(1 + \mu^2) \cos \Omega t \sin \Omega t - 2k\mu \sin^2 \Omega t,$$

$$\Delta k_{21}(t) = 2k\mu \cos^2 \Omega t + 2k \cos \Omega t \sin \Omega t - 2 \frac{N_0}{r}(1 + \mu^2) \cos \Omega t \sin \Omega t,$$

$$\Delta k_{22}(t) = 2 \frac{N_0}{r}(1 + \mu^2) \cos^2 \Omega t - 2k \cos^2 \Omega t + k\mu \sin 2\Omega t$$

and are a special case of equations (2.4a) if the terms Δk_{ij} are interpreted as perturbations. The analysis of Chap. 2 shows that asymmetric bearing stiffnesses have a stabilizing effect. This remains true also if we drop the assumption that the mass m of the pins is zero, which is more realistic for applications and is easily verified by numerical integration as can be seen in Fig. 4.3. The results were calculated using the parameters

$$c_x = 10^8 \text{ N/m}, \quad c_y = 2 \cdot 10^8 \text{ N/m}, \quad N_0 = 2 \text{ N}, \quad k = 10^6 \text{ N/m}, \quad M = 1 \text{ kg}$$
$$\mu = 0.8, \quad\quad\quad \Omega = 500 \text{ 1/s}, \quad\quad\quad r = 0.1 \text{ m}, \quad m = 0.2 \text{ kg}$$

for the asymmetric case. For the symmetric case $c_x = c_y = 10^8 \text{ N/m}$ was used.

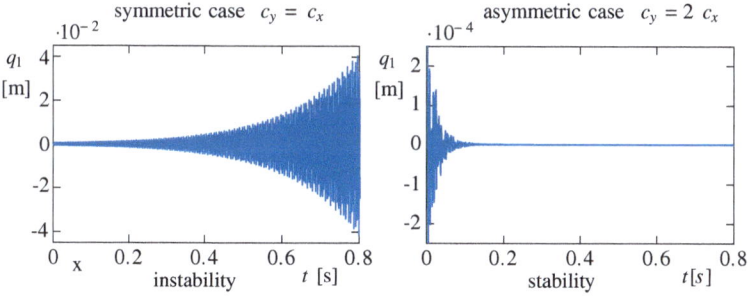

Fig. 4.3 Numerical integration of the equations of motion

Summarizing, it can be noted, that for the simple example studied, an easy design
modification can be proposed such that the system is stabilized. If a slight damping
is considered in the bearings, the results also carry over to the nonlinear case, since
asymptotic stability can be observed from the linear equations. In continuous systems
the choice of design modifications is not so obvious, therefore this case is addressed
in Chap. 6.

4.3 Asymmetric Wobbling Disk Model

In [91] a minimal model for brake squeal involving a rotating wobbling disk was
introduced in order to explain the excitation mechanism of brake squeal in disk
brakes. The study was based on solving an eigenvalue problem for linear equations
of motion with constant coefficients. In the sequel we slightly modify the model to
allow for asymmetric brake rotors and introduce nonlinear restoring forces arising
from a possible nonlinear material characteristic of the friction material [25].

4.3.1 Modeling Assumptions and Equations of Motion

The model shown in Fig. 4.4 consists of a rigid disk rotating about the d_3 axis at
constant angular velocity Ω corresponding to the nonholonomic constraint

$$^N\boldsymbol{\omega}^D \cdot \boldsymbol{d}_3 = \Omega. \tag{4.11}$$

For the linearized equations of motion this constraint is equivalent to $^N\boldsymbol{\omega}^D \cdot \boldsymbol{n}_3 = \Omega$
used in [91]. The rotor is free to tilt with respect to the $\boldsymbol{n}_1$–$\boldsymbol{n}_2$ plane. The torque due

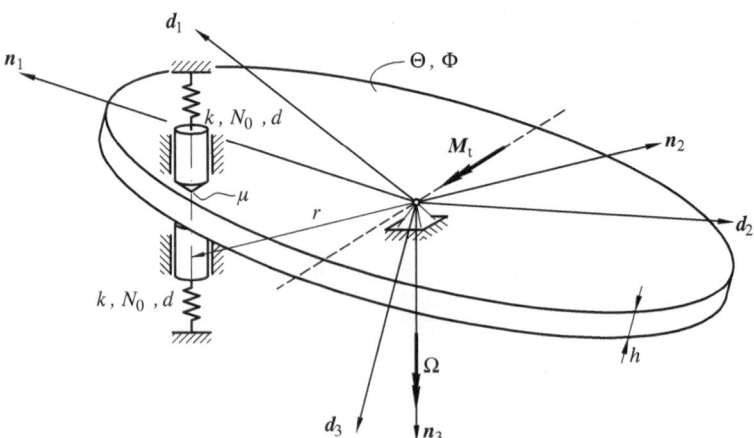

Fig. 4.4 Disk brake model with wobbling disk

to the mounting

$$M_t = M_{tk} + M_{td} \tag{4.12}$$

acting on the disk includes an elastic restoring torque

$$M_{tk} = -k_t \left(\arccos(n_3 \cdot d_2) - \frac{\pi}{2} \right) d_1 - (1 + \kappa_t) k_t \left(\arccos(n_3 \cdot d_1) - \frac{\pi}{2} \right) d_2 \tag{4.13}$$

proportional to the tilting angle between the two axes d_1, d_2 and the normal vector n_3, as well as a viscous damping torque

$$M_{td} = -d_t (^N \omega^D \cdot d_1) d_1 - (1 + \delta_t) d_t (^N \omega^D \cdot d_2) d_2 \tag{4.14}$$

proportional to the components of the angular velocity in the directions d_1 and d_2. This definition corresponds to a point-symmetric visco-elastic bedding of the disk with respect to the n_3 axis for $\kappa_t = \delta_t = 0$ and agrees with the formulation in [91] in the linear terms. The disk is in contact with two idealized massless brake pads which are pressed onto the surface of the disk by prestressed spring-damper elements (stiffness k, damping d, prestress N_0). Between the disk and the pad COULOMB friction occurs. If the generalized coordinates describing the orientation of the disk are defined as in [91], i.e. a CARDAN 1-2-3-rotation with angles q_1, q_2, q_3 and subsequent elimination of $\dot{q}_3$ using the nonholonomic constraint (4.11), the linearized equations of motion read

$$
\begin{bmatrix} \Theta & 0 \\ 0 & \Theta \end{bmatrix} \begin{bmatrix} \ddot{q}_1 \\ \ddot{q}_2 \end{bmatrix} + \left(\begin{bmatrix} d_t + 2dr^2 + \frac{1}{2}\mu N_0 \frac{h^2}{r\Omega} & \Phi\Omega \\ -\Phi\Omega - dh\mu r & (1+\delta_t)d_t \end{bmatrix} - \delta_t d_t \begin{bmatrix} -\sin^2 \Omega t & \sin \Omega t \cos \Omega t \\ \sin \Omega t \cos \Omega t & \sin^2 \Omega t \end{bmatrix} \right)
$$
$$
\begin{bmatrix} \dot{q}_1 \\ \dot{q}_2 \end{bmatrix} + \left(\begin{bmatrix} k_t + 2kr^2 + N_0 h & \frac{1}{2}\mu N_0 \frac{h^2}{r} \\ -\mu(khr + 2N_0 r) & (1+\kappa_t)k_t + (1+\mu^2)N_0 h \end{bmatrix} \right.
$$
$$
\left. -\kappa_t k_t \begin{bmatrix} -\sin^2 \Omega t & \sin \Omega t \cos \Omega t \\ \sin \Omega t \cos \Omega t & \sin^2 \Omega t \end{bmatrix} \right) \begin{bmatrix} q_1 \\ q_2 \end{bmatrix} = \begin{bmatrix} 0 \\ 0 \end{bmatrix}. \tag{4.15}
$$

Due to the asymmetry of the mounting, the equations of motion feature periodic coefficients. If the mounting is point symmetric (i.e. $\kappa_t = \delta_t = 0$), the linearized equations of motion simplify to the ones stated in [91]. For the following analysis the parameters

$$h = 0.02 \, \text{m}, \qquad r = 0.13 \, \text{m}, \qquad \Theta = 0.16 \, \text{kg/m}^2, \qquad \Phi = 2\Theta,$$
$$k_t = 1.88 \cdot 10^7 \, \text{Nm}, \quad k = 6.00 \cdot 10^6 \, \text{N/m}, \quad N_0 = 3.00 \, \text{kN}, \qquad \mu = 0.6,$$
$$d = 5 \, \text{Ns/m}, \qquad d_t = 0.1 \, \text{Nms}$$

taken from [91] will be used. The dimensionless values κ_t, δ_t characterizing the asymmetry and the rotational speed Ω will be varied.

4.3.2 Formulation of a Perturbation Problem

Changing the sequence of the CARDAN rotations, a formulation of the equations of motion more suitable for analytical investigations is obtained. Defining the generalized coordinates by a CARDAN 3-1-2-rotation with angles q_3, q_1, q_2 and subsequent elimination of $\dot{q}_3$ using the nonholonomic constraint (4.11) yields the linearized equations of motion

$$\begin{bmatrix} \Theta & 0 \\ 0 & \Theta \end{bmatrix} \begin{bmatrix} \ddot{q}_1 \\ \ddot{q}_2 \end{bmatrix} + \begin{bmatrix} \frac{1}{2}\mu N_0 \frac{h^2}{r\Omega} \cos^2 \Omega t & -\frac{1}{2}\mu N_0 \frac{h^2}{r\Omega} \sin \Omega t \cos \Omega t \\ -\frac{1}{2}\mu N_0 \frac{h^2}{r\Omega} \sin \Omega t \cos \Omega t & \frac{1}{2}\mu N_0 \frac{h^2}{r\Omega} \sin^2 \Omega t \end{bmatrix} \begin{bmatrix} \dot{q}_1 \\ \dot{q}_2 \end{bmatrix}$$
$$+ \begin{bmatrix} k_{11} & k_{12} \\ k_{21} & k_{22} \end{bmatrix} \begin{bmatrix} q_1 \\ q_2 \end{bmatrix} = \begin{bmatrix} 0 \\ 0 \end{bmatrix},$$
$$(4.16)$$

with

$$k_{11} = k_{t1} + N_0 h + \Theta\Omega^2 + 2kr^2 \cos^2 \Omega t + (2\mu N_0 r + h\mu kr) \sin \Omega t \cos \Omega t - h\mu^2 N_0 \sin^2 \Omega t,$$
$$k_{12} = h\mu kr \sin^2 \Omega t + h\mu^2 N_0 \sin \Omega t \cos \Omega t - 2\mu N_0 r \cos^2 \Omega t - 2kr^2 \sin \Omega t \cos \Omega t,$$
$$k_{21} = -2kr^2 \cos \Omega t \sin \Omega t - 2\mu N_0 r \cos^2 \Omega t - h\mu kr \cos^2 \Omega t + h\mu^2 N_0 \sin \Omega t \cos \Omega t,$$
$$k_{22} = k_{t2} + N_0 h + \Theta\Omega^2 + 2kr^2 \sin^2 \Omega t + (2\mu N_0 r + h\mu kr) \sin \Omega t \cos \Omega t - h\mu^2 N_0 \cos^2 \Omega t$$

stated here for the undamped case only, for the sake of brevity. Introducing damping in the pad and the damping torque (4.14) yields additional terms in the matrices proportional to $q_{1,2}$ and $\dot{q}_{1,2}$. The equations of motion now have the form

$$M\ddot{q} + \varepsilon \Delta D(t)\dot{q} + (K + \varepsilon \Delta K(t))q = 0, \qquad (4.17)$$

that has been analyzed in Chap. 2. The matrices $M = M^{\mathrm{T}}$ and $K = K^{\mathrm{T}}$ are the mass and the stiffness matrix of the undamped disk without pads. The terms occurring from the pads and the damping torque are interpreted as perturbations and represented by the matrices $\Delta D(t) = \Delta D(t + T)$, $\Delta K(t) = \Delta K(t + T)$, which are periodic with respect to $T = 2\pi/\Omega$. The parameter $\varepsilon \ll 1$ can be interpreted as a norm of multiple parameters depending linearly on ε and vanishing for $\varepsilon = 0$. In Chap. 2 it has been shown that splitting up the eigenfrequencies of the unperturbed problem of (4.17) helps to stabilize the system. This is done by setting $\delta_t = 0$ and varying κ_t. Figure 4.5 shows the critical speed, i.e. the speed at which the trivial solution loses stability, as a function of κ_t. It can be clearly seen that the critical speed is minimal in the symmetric case ($\kappa_t = 0$), and increases significantly in the nonsymmetric case for $|\kappa_t| > 1$ ‰. The stability boundary in Fig. 4.5 is in good agreement with the analytical approximation of the corresponding general two dimensional case in [73].

Fig. 4.5 Critical speed (speed at which the trivial solution loses stability) for varying κ_t

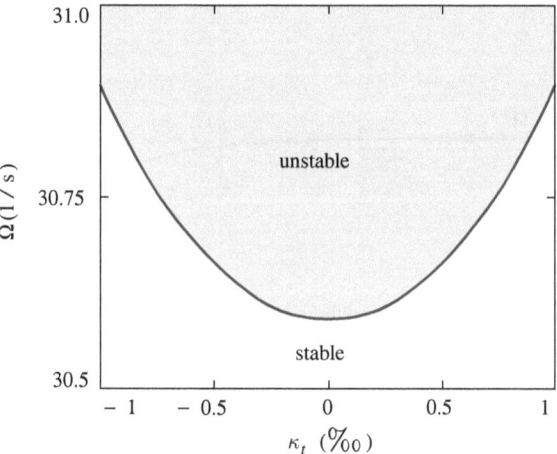

4.4 Rotor with Translational and Tilting Degrees of Freedom

The systems studied in the previous two sections could be presented in the same mathematical format so that we were able to apply the results of Chap. 2 for their passive stabilization by construction measures. However, we note that the modifications made were quite different. The rotor studied in Sect. 4.2 had two translational degrees of freedom, the equations of could be written in the form (4.17) in nonrotating coordinates and the system could be stabilized by introducing a nonrotating asymmetry in the bearings. In rotating coordinates, on the other hand, the equations of motion had gyroscopic terms, which weakens the reasoning from Chap. 2.

The equations of motion for the wobbling disk model discussed in Sect. 4.3 could also be brought into the form (4.17), however, here the equations of motion were set up with respect to a frame rotating with the disk. The equations of motion, where the degrees of freedom were defined with respect to a nonrotating frame, featured gyroscopic terms.

We conclude from the previous two examples that, in order to stabilize translational degrees of freedom of the rotor, a nonrotating asymmetry in the bearings was required, whereas for the tilting degrees of freedom, a rotating asymmetry had to be introduced.

In the following we will investigate what happens if a rotor contains translational and tilting degrees of freedom. Consider a shaft rotating with constant angular velocity Ω, shown in Fig. 4.6, consisting of a massless beam with bending stiffnesses EI_1 and EI_2 about it's principal axes and a thin rigid disk (mass m and radius r). In addition, we consider a nonrotating elastic coupling to the ground, represented by springs with stiffnesses k_1 and k_2. In the following we use a coordinate frame rotating with the disk spanned by d_1, d_2, d_3 and the inertial frame spanned by n_1, n_2, n_3. Motivated by the dry friction examples in the previous section, we assume

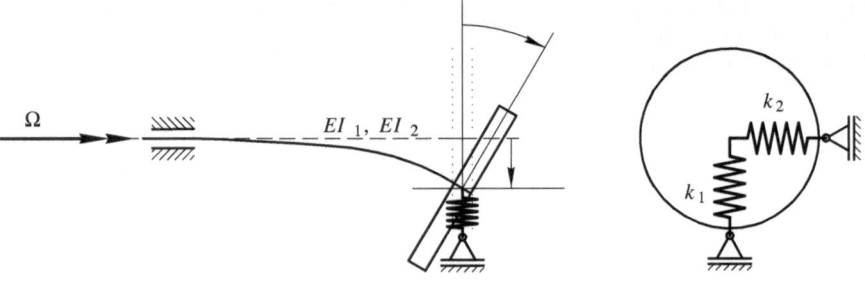

Fig. 4.6 LAVAL rotor with translational and tilting degrees of freedom

linear nonconservative forces acting on the disc, from either rotating contacts

$$F_n = n_t[(r \cdot d_1)d_2 - (r \cdot d_2)d_1], \tag{4.18a}$$

$$M_n = n_r[((d_3 \times n_3) \cdot d_1)d_2 - ((d_3 \times n_3) \cdot d_2)d_1] \tag{4.18b}$$

or nonrotating contacts as

$$F_n = n_t[(r \cdot n_1)n_2 - (r \cdot n_2)n_1], \tag{4.19a}$$

$$M_n = n_r[((d_3 \times n_3) \cdot n_1)n_2 - ((d_3 \times n_3) \cdot n_2)n_1]. \tag{4.19b}$$

The formulation of the forcing is such that an elongation in one direction causes a force into the perpendicular direction and a rotation about one axis causes a torque about the other. The magnitude of the forcing for translational and rotational forcing is n_t and n_r, respectively. In order to keep the equations short in the following we neglect inner and outer damping. The disk is given two translational degrees of freedom q_1, q_2 and two tilting angles q_3, q_4 which each can be defined either with respect to rotating or nonrotating coordinate systems. The linearized equations of motion

$$M\ddot{q} + (G(t))\dot{q} + (K(t) + N(t))q = 0, \tag{4.20}$$

have a different structure, depending on the way the coordinates are defined. In the following we will investigate different settings:

(a) translatory and tilting dof are defined with respect to a rotating frame,
(b) translatory and tilting dof are defined with respect to a stationary frame,
(c) translatory dof are defined in a rotating frame and tilting dof with respect to a stationary frame,
(d) translatory dof are defined in a stationary frame and tilting dof with respect to a rotating frame

by writing the equations for each possible combination. In all of the settings the mass matrix reads

$$M = \begin{bmatrix} m & 0 & 0 & 0 \\ 0 & m & 0 & 0 \\ 0 & 0 & \frac{1}{4}m\,r^2 & 0 \\ 0 & 0 & 0 & \frac{1}{4}m\,r^2 \end{bmatrix}, \tag{4.21}$$

and the matrix due to the nonconservative forces reads

$$N = \begin{bmatrix} 0 & n_t & 0 & 0 \\ -n_t & 0 & 0 & 0 \\ 0 & 0 & 0 & n_r \\ 0 & 0 & -n_r & 0 \end{bmatrix}, \tag{4.22}$$

since the upper an the lower block are special cases of the rotation matrix multiplied by a factor and

$$\begin{bmatrix} \cos\varphi & \sin\varphi \\ -\sin\varphi & \cos\varphi \end{bmatrix}^T \begin{bmatrix} \cos\psi & \sin\psi \\ -\sin\psi & \cos\psi \end{bmatrix} \begin{bmatrix} \cos\varphi & \sin\varphi \\ -\sin\varphi & \cos\varphi \end{bmatrix} = \begin{bmatrix} \cos\psi & \sin\psi \\ -\sin\psi & \cos\psi \end{bmatrix} \tag{4.23}$$

for all φ, ψ. The matrix entries k_{ij} of the matrix $K(t)$ and the entries g_{ij} of $G(t)$ are different in every setting. The shortest form of equations of motion is obtained for case (a), where we define both translatory and tilting degrees of freedom with respect to rotating coordinates. The matrix proportional to the generalized velocities then reads

$$G = \begin{bmatrix} 0 & -2\,m\,\Omega & 0 & 0 \\ 2\,m\,\Omega & 0 & 0 & 0 \\ 0 & 0 & 0 & 0 \\ 0 & 0 & 0 & 0 \end{bmatrix}.$$

We observe that we have gyroscopic terms with respect to the translational degrees of freedom which can be explained by CORIOLIS forces. The entries of the stiffness matrix are given by

$$k_{11} = k_1 + \frac{12\,E\,I_1}{L^3} - m\,\Omega^2 - (k_1 - k_2)\sin(\Omega t)^2,$$

$$k_{12} = n_t - (k1 - k_2)\sin(\Omega t)\cos(\Omega t),$$

$$k_{13} = 0,$$

$$k_{14} = \frac{6\,E\,I_1}{L^2},$$

$$k_{21} = -(k_1 - k_2)\sin(\Omega t)\cos(\Omega t) - n_t,$$

$$k_{22} = k_2 + \frac{12\,E\,I_2}{L^3} + (k_1 - k_2)\sin(\Omega t)^2 - m\,\Omega^2,$$

$$k_{23} = -\frac{6\,E\,I_2}{L^2},$$

$$k_{24} = 0,$$

$$k_{31} = 0,$$

$$k_{32} = -\frac{6\,E\,I_2}{L^2},$$

$$k_{33} = \frac{4\,E\,I_2}{L} + \frac{1}{4}\,m\,\Omega^2\,r^2,$$

$$k_{34} = -n_r,$$

$$k_{41} = \frac{6\,E\,I_1}{L^2},$$

$$k_{42} = 0,$$

$$k_{43} = n_r,$$

$$k_{44} = \frac{4\,E\,I_1}{L} + \frac{1}{4}\,m\,\Omega^2\,r^2,$$

where the only time dependent terms occur due to the asymmetric bedding $k_1 - k_2 \neq 0$. If according to case (b) we define both translations and tilting in nonrotating coordinates we obtain

$$G = \begin{bmatrix} 0 & 0 & 0 & 0 \\ 0 & 0 & 0 & 0 \\ 0 & 0 & 0 & \frac{1}{2}\,m\,\Omega\,r^2 \\ 0 & 0 & -\frac{1}{2}\,m\,\Omega\,r^2 & 0 \end{bmatrix},$$

where the gyroscopic terms, in contrast to the previous case, occur in the tilting degrees of freedom. This time the occurrence of the gyroscopic terms cannot be explained by CORIOLIS forces. The entries of the stiffness matrix read

$$k_{11} = k_1 + \frac{12\,(\frac{E I_1}{L^3} - (\frac{E I_1}{L^3} - \frac{E I_2}{L^3})\,\sin(\Omega t)^2)}{L^3},$$

$$k_{12} = n_t + \frac{12\,(\frac{E I_1}{L^3} - \frac{E I_2}{L^3})\,\sin(\Omega t)\,\cos(\Omega t)}{L^3},$$

$$k_{13} = -\frac{6\,(E I_1 - E I_2)\,\sin(\Omega t)\,\cos(\Omega t)}{L^2},$$

$$k_{14} = \frac{6\,(E I_1 - (E I_1 - E I_2)\,\sin(\Omega t)^2)}{L^2},$$

$$k_{21} = \frac{12\,(E I_1 - E I_2)\,\sin(\Omega t)\,\cos(\Omega t)}{L^3} - n_t,$$

$$k_{22} = k_2 + \frac{12\,(E I_2 + (E I_1 - E I_2)\,\sin(\Omega t)^2)}{L^3},$$

$$k_{23} = -\frac{6\,(E I_2 + (E I_1 - E I_2)\,\sin(\Omega t)^2)}{L^2},$$

$$k_{24} = \frac{6\,(E I_1 - E I_2)\,\sin(\Omega t)\,\cos(\Omega t)}{L^2},$$

$$k_{31} = -\frac{6\,(E I_1 - E I_2)\,\sin(\Omega t)\,\cos(\Omega t)}{L^2},$$

$$k_{32} = -\frac{6\,(E I_2 + (E I_1 - E I_2)\,\sin(\Omega t)^2)}{L^2},$$

$$k_{33} = \frac{4\,(E I_2 + (E I_1 - E I_2)\,\sin(\Omega t)^2)}{L},$$

$$k_{34} = -n_r - \frac{4\,(E I_1 - E I_2)\,\sin(\Omega t)\,\cos(\Omega t)}{L},$$

$$k_{41} = \frac{6\,(E I_1 - (E I_1 - E I_2)\,\sin(\Omega t)^2)}{L^2},$$

$$k_{42} = \frac{6\,(E I_1 - E I_2)\,\sin(\Omega t)\,\cos(\Omega t)}{L^2},$$

$$k_{43} = n_r - \frac{4\,(E I_1 - E I_2)\,\sin(\Omega t)\,\cos(\Omega t)}{L},$$

$$k_{44} = \frac{4\,(E I_1 - (E I_1 - E I_2)\,\sin(\Omega t)^2)}{L}$$

and the periodic terms now originate from the difference in the bending stiffness, i.e. $E I_1 \neq E I_2$. Defining the translational degrees of freedom in rotating coordinates and the tilting ones in spatially fixed ones, as in case (c), yields

$$G = \begin{bmatrix} 0 & -2\,m\,\Omega & 0 & 0 \\ 2\,m\,\Omega & 0 & 0 & 0 \\ 0 & 0 & 0 & \frac{1}{2}\,m\,\Omega\,r^2 \\ 0 & 0 & -\frac{1}{2}\,m\,\Omega\,r^2 & 0 \end{bmatrix},$$

where gyroscopic terms appear in both tilting and translational degrees of freedom. The entries of the stiffness matrix read

$$k_{11} = k_1 + \frac{12\,E I_1}{L^3} - m\,\Omega^2 - (k_1 - k_2)\,\sin(\Omega t)^2,$$

$$k_{12} = n_t - (k_1 - k_2)\,\sin(\Omega t)\,\cos(\Omega t),$$

$$k_{13} = -\frac{6\,E I_1\,\sin(\Omega t)}{L^2},$$

$$k_{14} = \frac{6\,E I_1\,\cos(\Omega t)}{L^2},$$

$$k_{21} = -(k_1 - k_2)\,\sin(\Omega t)\,\cos(\Omega t) - n_t,$$

$$k_{22} = k_2 + \frac{12\,E I_2}{L^3} + (k_1 - k_2)\,\sin(\Omega t)^2 - m\,\Omega^2,$$

$$k_{23} = -\frac{6\,E I_2\,\cos(\Omega t)}{L^2},$$

$$k_{24} = -\frac{6\,E I_2\,\sin(\Omega t)}{L^2},$$

$$k_{31} = -\frac{6\,EI_1\,\sin(\Omega t)}{L^2},$$

$$k_{32} = -\frac{6\,EI_2\,\cos(\Omega t)}{L^2},$$

$$k_{33} = \frac{4\,(EI_2 + (EI_1 - EI_2)\,\sin(\Omega t)^2)}{L},$$

$$k_{34} = -n_r - \frac{4\,(EI_1 - EI_2)\,\sin(\Omega t)\,\cos(\Omega t)}{L},$$

$$k_{41} = \frac{6\,EI_1\,\cos(\Omega t)}{L^2},$$

$$k_{42} = -\frac{6\,EI_2\,\sin(\Omega t)}{L^2},$$

$$k_{43} = n_r - \frac{4\,(EI_1 - EI_2)\,\sin(\Omega t)\,\cos(\Omega t)}{L},$$

$$k_{44} = \frac{4\,(EI_1 - (EI_1 - EI_2)\,\sin(\Omega t)^2)}{L}$$

and the time dependent terms originate from both $k_1 \neq k_2$ and $EI_1 \neq EI_2$. Finally, in case (d), defining the translatorial degrees of freedom with respect to nonrotating coordinates and the tilting ones in a rotating frame, we obtain

$$G = \begin{bmatrix} 0 & 0 & 0 & 0 \\ 0 & 0 & 0 & 0 \\ 0 & 0 & 0 & 0 \\ 0 & 0 & 0 & 0 \end{bmatrix},$$

which means that no gyroscopic terms occur. The stiffness matrix reads

$$k_{11} = k_1 + \frac{12\,(EI_1 - (EI_1 - EI_2)\,\sin(\Omega t)^2)}{L^3},$$

$$k_{12} = n_t + \frac{12\,(EI_1 - EI_2)\,\sin(\Omega t)\,\cos(\Omega t)}{L^3},$$

$$k_{13} = \frac{6\,EI_2\,\sin(\Omega t)}{L^2},$$

$$k_{14} = \frac{6\,EI_1\,\cos(\Omega t)}{L^2},$$

$$k_{21} = \frac{12\,(EI_1 - EI_2)\,\sin(\Omega t)\,\cos(\Omega t)}{L^3} - n_t,$$

$$k_{22} = k_2 + \frac{12\,(EI_2 + (EI_1 - EI_2)\,\sin(\Omega t)^2)}{L^3},$$

$$k_{23} = -\frac{6\,EI_2\,\cos(\Omega t)}{L^2},$$

$$k_{24} = \frac{6\,EI_1\,\sin(\Omega t)}{L^2},$$

$$k_{31} = \frac{6\,E I_2\,\sin(\Omega t)}{L^2},$$

$$k_{32} = -\frac{6\,E I_2\,\cos(\Omega t)}{L^2},$$

$$k_{33} = \frac{4\,E I_2}{L} + \frac{1}{4}\,m\,\Omega^2\,r^2,$$

$$k_{34} = -n_r,$$

$$k_{41} = \frac{6\,E I_1\,\sin(\Omega t)}{L^2},$$

$$k_{42} = \frac{6\,E I_1\,\sin(\Omega t)}{L^2},$$

$$k_{43} = n_r,$$

$$k_{44} = \frac{4\,E I_1}{L} + \frac{1}{4}\,m\,\Omega^2\,r^2.$$

Summarizing, in the different settings, the equations of motion have the following structure:

(a) translatory and tilting dof are defined with respect to a rotating frame,

$$\begin{bmatrix} M_{tt} & 0 \\ 0 & M_{rr} \end{bmatrix}\ddot{q} + \begin{bmatrix} G_{tt} & 0 \\ 0 & 0 \end{bmatrix}\dot{q} + \begin{bmatrix} K_{tt}(t)+N_{tt} & K_{tr} \\ K_{tr} & K_{rr}+N_{rr} \end{bmatrix}q = 0 \quad (4.24a)$$

(b) translatory and tilting dof are defined with respect to a stationary frame,

$$\begin{bmatrix} M_{tt} & 0 \\ 0 & M_{rr} \end{bmatrix}\ddot{q} + \begin{bmatrix} 0 & 0 \\ 0 & G_{rr} \end{bmatrix}\dot{q} + \begin{bmatrix} K_{tt}(t)+N_{tt} & K_{tr}(t) \\ K_{tr}(t) & K_{rr}(t)+N_{rr} \end{bmatrix}q = 0 \quad (4.24b)$$

(c) translatory dof are defined in a rotating frame and tilting dof with respect to a stationary frame,

$$\begin{bmatrix} M_{tt} & 0 \\ 0 & M_{rr} \end{bmatrix}\ddot{q} + \begin{bmatrix} G_{tt} & 0 \\ 0 & G_{rr} \end{bmatrix}\dot{q} + \begin{bmatrix} K_{tt}(t)+N_{tt} & K_{tr}(t) \\ K_{tr}(t) & K_{rr}(t)+N_{rr} \end{bmatrix}q = 0$$
$$(4.24c)$$

(d) translatory dof are defined in a stationary frame and tilting dof with respect to a rotating frame

$$\begin{bmatrix} M_{tt} & 0 \\ 0 & M_{rr} \end{bmatrix}\ddot{q} + \begin{bmatrix} 0 & 0 \\ 0 & 0 \end{bmatrix}\dot{q} + \begin{bmatrix} K_{tt}(t)+N_{tt} & K_{tr}(t) \\ K_{tr}(t) & K_{rr}+N_{rr} \end{bmatrix}q = 0 \quad (4.24d)$$

We observe that, although the skew symmetric terms in the stiffness matrix considered in this case are constant, in none of the cases considered above, we obtain a system of equations with constant coefficients, since either the difference of $E I_1$ and $E I_2$ or the difference of k_1 and k_2 yield periodic coefficients. Therefore for the general case it is not straightforward to form an unperturbed problem so that the findings of Chap. 2 are applicable. Nevertheless, for small angular velocities the gyroscopic terms are small and the periodic terms depend on a few parameters which in some cases will be small. If for example $k_1 - k_2$ is small and the same is true for the angular velocity, formulation (a) is most suitable to use the results of Chap. 2, where the gyroscopic terms and the time periodic terms can be seen as perturbations. Since gyroscopic and nonconservative terms are present in the equations, a stabilization can of course only be expected in the presence of damping. However, a splitting of the eigenfrequencies can be seen to have a stabilizing effect.

4.5 Elimination of Gyroscopic Terms in an Arbitrary Multibody System

In the examples of the last section gyroscopic terms could be avoided by formulating the equations of motion with respect to rotating coordinate systems. In this section we will study the elimination of the gyroscopic terms from a mathematical point of view and show that it can be performed for arbitrary rigid body systems. Consider the general case of a gyroscopic two degree of freedom system

$$\begin{bmatrix} m_1 & 0 \\ 0 & m_2 \end{bmatrix} \begin{bmatrix} \ddot{q}_1 \\ \ddot{q}_2 \end{bmatrix} + \left(\begin{bmatrix} 0 & g \\ -g & 0 \end{bmatrix} + \begin{bmatrix} d_{11} & d_{12} \\ d_{12} & d_{22} \end{bmatrix} \right) \begin{bmatrix} \dot{q}_1 \\ \dot{q}_2 \end{bmatrix} + \begin{bmatrix} k_{11} & k_{12} \\ k_{21} & k_{22} \end{bmatrix} \begin{bmatrix} q_1 \\ q_2 \end{bmatrix} = \begin{bmatrix} 0 \\ 0 \end{bmatrix},$$
(4.25)

which, using $q = M^{\frac{1}{2}}x$, can be transformed into the form

$$\begin{bmatrix} 1 & 0 \\ 0 & 1 \end{bmatrix} \begin{bmatrix} \ddot{x}_1 \\ \ddot{x}_2 \end{bmatrix} + \left(\begin{bmatrix} 0 & \tilde{g} \\ -\tilde{g} & 0 \end{bmatrix} + \begin{bmatrix} \tilde{d}_{11} & \tilde{d}_{12} \\ \tilde{d}_{12} & \tilde{d}_{22} \end{bmatrix} \right) \begin{bmatrix} \dot{x}_1 \\ \dot{x}_2 \end{bmatrix} + \begin{bmatrix} \tilde{k}_{11} & \tilde{k}_{12} \\ \tilde{k}_{21} & \tilde{k}_{22} \end{bmatrix} \begin{bmatrix} x_1 \\ x_2 \end{bmatrix} = \begin{bmatrix} 0 \\ 0 \end{bmatrix}.$$
(4.26)

The transformation

$$\begin{bmatrix} y_1 \\ y_2 \end{bmatrix} = \begin{bmatrix} \cos(\frac{\tilde{g}}{2}t) & \sin(\frac{\tilde{g}}{2}t) \\ -\sin(\frac{\tilde{g}}{2}t) & \cos(\frac{\tilde{g}}{2}t) \end{bmatrix} \begin{bmatrix} x_1 \\ x_2 \end{bmatrix},$$
(4.27)

eliminates the gyroscopic terms and yields

$$\begin{bmatrix} 1 & 0 \\ 0 & 1 \end{bmatrix} \begin{bmatrix} \ddot{x}_1 \\ \ddot{x}_2 \end{bmatrix} + \begin{bmatrix} \bar{d}_{11}(t) & \bar{d}_{12}(t) \\ \bar{d}_{12}(t) & \bar{d}_{22}(t) \end{bmatrix} \begin{bmatrix} \dot{x}_1 \\ \dot{x}_2 \end{bmatrix} + \begin{bmatrix} \bar{k}_{11}(t) & \bar{k}_{12}(t) \\ \bar{k}_{21}(t) & \bar{k}_{22}(t) \end{bmatrix} \begin{bmatrix} x_1 \\ x_2 \end{bmatrix} = \begin{bmatrix} 0 \\ 0 \end{bmatrix},$$ (4.28)

where the matrices depend on time. For a general linear system of rigid bodies the equations of motion can be formulated such that the gyroscopic matrix has anti-diagonal form. To achieve this form, define the 3 translational dof along spatially fixed axes and all rotations about body fixed axes. If the body rotates, define one of the tilting dof around the axis of rotation and the two remaining dof with respect to axis rotating with the body and perpendicular to the axes of rotation. The only remaining time dependent terms come from inertially fixed couplings and yield symmetric matrices. Performing the same transformation as for the two dof system for each pair of coordinates which is coupled through the gyroscopic terms one can transform

$$M\ddot{q} + (G + D(t))\dot{q} + K(t)q = 0, \tag{4.29}$$

where M is diagonal and G is skew-symmetric and anti-diagonal into

$$I\ddot{q} + D(t)\dot{q} + K(t)q = 0, \tag{4.30}$$

were the entries of $K(t)$ are each periodic with respect to different potentially incommensurable periods. This case is also referred to as quasi periodic and the stability of the trivial solution is difficult to investigate since FLOQUET theory is not applicable in this case.

4.6 Summary

In this chapter we investigated simple discrete models of rotors in frictional contact. The systems serve as examples for systems in which self-excited vibrations can be avoided by splitting up multiple eigenfrequencies. We discussed the influence of different choices of generalized coordinates on the equations of motion and their stability behavior and the sources and avoidance of gyroscopic terms in the equations of motion.

Chapter 5
Passive Stabilization in Continuous Systems

In the last chapter we have seen simple illustrative examples for systems which could be designed robust against self-excited vibrations through structural modifications. We will see that a lot of the results obtained in the last chapter carry over to the continuous case as well. At first we study a quite general problem of a shell in frictional contact and then discuss a special example of a beam with periodic boundary conditions.

5.1 Shells in Frictional Contact

In this section we consider shells in frictional contact. This setting can be used to model the singing of wine glasses, with rotating or nonrotating glass. Analytical models for disc brake squeal can be seen as a special case of the problem treated.

We consider a shell in contact with an idealized friction pad pressed upon the shell by a prestressed spring. Two settings will be investigated, first we consider a shell rotating at constant angular velocity, and second a shell subjected to a rotating pad. We assume that the shell consists of linear elastic isotropic material, that planar crossections stay planar and perpendicular to the neutral surface of the shell at all times, and that the transverse stress σ_{33} is negligible (LOVE simplification). The mass of each crossection is assumed to be concentrated on the neutral plane. The geometry of the shell is described by two orthogonal curvilinear coordinates. In material coordinates ξ, η a material point on the neutral surface of the shell is described by

$$\boldsymbol{p}_0 = f_1(\xi, \eta, t)\boldsymbol{e}_x + f_2(\xi, \eta, t)\boldsymbol{e}_y + f_3(\xi, \eta, t)\boldsymbol{e}_z \tag{5.1}$$

in a global coordinate system, where the explicit time dependence of the f_i originates from a prescribed motion of the undeformed shell. For a shell rotating about the $\boldsymbol{e}_z$ axis at constant angular velocity Ω, we can write

$$\boldsymbol{p}_0 = r(\xi, \eta)\cos(\Omega t + \eta)\boldsymbol{e}_x + r(\xi, \eta)\sin(\Omega t + \eta)\boldsymbol{e}_y + f_3(\xi, \eta)\boldsymbol{e}_z. \tag{5.2}$$

G. Spelsberg-Korspeter, *Robust Structural Design against Self-Excited Vibrations*, SpringerBriefs in Applied Sciences and Technology, DOI: 10.1007/978-3-642-36552-2_5, © The Author(s) 2013

If rotating shells or shells with moving loads are considered, it can be convenient to define the coordinates α, ϕ with respect to moving coordinate systems which in some cases can eliminate an explicit time dependence of the f_i in (5.2) (cf. Fig. 5.1). It has to be noted however, that, when considering time derivatives, the material coordinates have to be kept constant. Here we concentrate on shells which can easily be modeled by cylindrical coordinates. Three cases are of major interest in this book and the relation of α and ϕ on ξ and η is stated accordingly (cf. Fig. 5.1):

(a) rotating shell in nonrotating coordinates $\alpha = \xi$, $\phi = \eta + \Omega t$,
(b) rotating shell in rotating coordinates $\alpha = \xi$, $\phi = \eta$,
(c) nonrotating shell in rotating coordinates $\alpha = \xi$, $\phi = \eta - \Omega t$.

For case (a), the position vector of a point on the neutral surface reads

$$\boldsymbol{p}_0 = r(\alpha, \phi - \Omega t)\cos\phi \boldsymbol{e}_x + r(\alpha, \phi - \Omega t)\sin\phi \boldsymbol{e}_y + f_3(\alpha, \phi - \Omega t)\boldsymbol{e}_z. \quad (5.3)$$

In case (b) we obtain

$$\boldsymbol{p}_0 = r(\alpha, \phi)\cos(\Omega t + \phi)\boldsymbol{e}_x + r(\alpha, \phi)\sin(\Omega t + \phi)\boldsymbol{e}_y + f_3(\alpha, \phi)\boldsymbol{e}_z \quad (5.4)$$

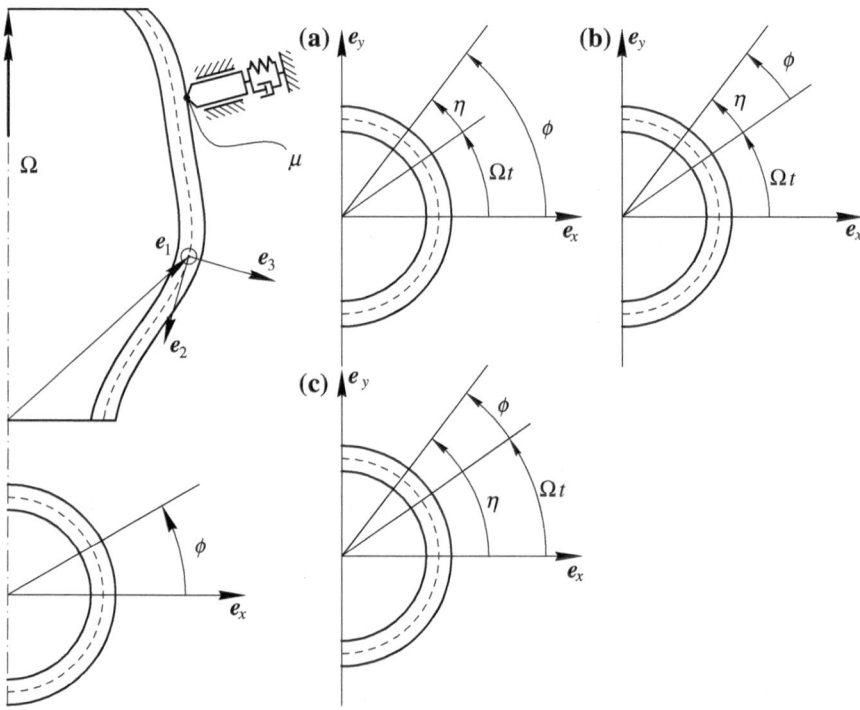

Fig. 5.1 Rotating frame in different reference frames

and for case (c) we have

$$\boldsymbol{p}_0 = r(\alpha, \phi + \Omega t) \cos(\Omega t + \phi)\boldsymbol{e}_x + r(\alpha, \phi + \Omega t) \sin(\Omega t + \phi)\boldsymbol{e}_y + f_3(\alpha, \phi + \Omega t)\boldsymbol{e}_z.$$

(5.5)

We note that for the case of a purely symmetric shell, $r(\alpha, \phi) = r$ is constant and $f_3(\alpha, \phi) = f_3(\alpha)$. The curvilinear coordinates define a local coordinate system given by

$$\boldsymbol{e}_1 = \frac{\boldsymbol{p}_{0,\alpha}}{\|\boldsymbol{p}_{0,\alpha}\|}, \quad \boldsymbol{e}_2 = \frac{\boldsymbol{p}_{0,\phi}}{\|\boldsymbol{p}_{0,\phi}\|}, \quad \boldsymbol{e}_3 = \boldsymbol{e}_1 \times \boldsymbol{e}_2.$$

(5.6)

We assume that points on the neutral surface of the shell have a degree of freedom in each direction of the corresponding local coordinate system, i.e. in the deformed configuration of the shell the position vector of a point on the neutral surface of the shell is given by

$$\boldsymbol{p}_M = \boldsymbol{p}_0 + u(\alpha, \phi)\boldsymbol{e}_1 + v(\alpha, \phi)\boldsymbol{e}_2 + w(\alpha, \phi)\boldsymbol{e}_3.$$

(5.7)

In order to obtain velocities of material points of the shell their corresponding position vectors have to be differentiated with respect to time with the material coordinates kept constant.

5.1.1 Principle of Virtual Work

Using the assumptions stated above, the principle of virtual work reads

$$\int_V \left[\rho \frac{\mathrm{d}^2}{\mathrm{d}t^2} \boldsymbol{p}_M \cdot \delta \boldsymbol{p}_M + \sum_{ij} \sigma_{ij} \delta \epsilon_{ij} \right] \mathrm{d}V = \sum_i F_i \cdot \delta \boldsymbol{p}_i.$$

(5.8)

In the following sections we derive and state the elastic restoring terms, the contribution of the contact forces to the virtual work and the inertia terms.

Inertia Terms

The inertia terms are obtained by differentiating (5.7) with respect to time keeping the material coordinates ξ, η constant. Depending on the choice of coordinates α, ϕ, the inertia terms take different forms.

Elastic Restoring Terms

The stress-strain relations are given by

$$\sigma_{11} = \frac{E}{1 - v^2}(\varepsilon_{11} + v\varepsilon_{22}), \quad \sigma_{22} = \frac{E}{1 - v^2}(\varepsilon_{22} + v\varepsilon_{11}), \quad \sigma_{21} = \sigma_{12} = \frac{E}{1 + v}\varepsilon_{12}$$
$$(5.9)$$

and the strain displacement relations using the LOVE simplifications read [71]

$$\varepsilon_{11} = \varepsilon_{11}^0 + \zeta k_{11}, \tag{5.10a}$$

$$\varepsilon_{22} = \varepsilon_{22}^0 + \zeta k_{22}, \tag{5.10b}$$

$$\varepsilon_{12} = \varepsilon_{12}^0 + \zeta k_{12}, \tag{5.10c}$$

where the quantities ε_{ij}^0 are membrane strains and ζk_{ij} indicate bending strains, which according to the assumptions stated, vary linearly through the thickness of the shell, i.e. ζ measures the distance of a point from the neutral plane in e_3 direction. The detailed expressions read [71]

$$\varepsilon_{11}^0 = \frac{1}{A_1}\frac{\partial u}{\partial \phi} + \frac{v}{A_1 A_2}\frac{\partial A_1}{\partial \alpha} + \frac{w}{R_1}, \tag{5.11a}$$

$$\varepsilon_{22}^0 = \frac{1}{A_2}\frac{\partial v}{\partial \alpha} + \frac{u}{A_1 A_2}\frac{\partial A_2}{\partial \phi} + \frac{w}{R_2}, \tag{5.11b}$$

$$\varepsilon_{12}^0 = \frac{A_2}{A_1}\frac{\partial}{\partial \phi}\left(\frac{v}{A_2}\right) + \frac{A_1}{A_2}\frac{\partial}{\partial \alpha}\left(\frac{u}{A_1}\right), \tag{5.11c}$$

and

$$k_{11} = \frac{1}{A_1}\frac{\partial \beta_1}{\partial \phi} + \frac{\beta_2}{A_1 A_2}\frac{\partial A_1}{\partial \alpha}, \tag{5.12a}$$

$$k_{22} = \frac{1}{A_2}\frac{\partial \beta_2}{\partial \alpha} + \frac{\beta_1}{A_1 A_2}\frac{\partial A_2}{\partial \phi}, \tag{5.12b}$$

$$k_{12} = \frac{A_2}{A_1}\frac{\partial}{\partial \alpha}\left(\frac{\beta_2}{A_2}\right) + \frac{A_1}{A_2}\frac{\partial}{\partial \alpha}\left(\frac{\beta_1}{A_1}\right), \tag{5.12c}$$

with

$$\beta_1 = \frac{u}{R_1} - \frac{1}{A_1}\frac{\partial w}{\partial \alpha}, \tag{5.13a}$$

$$\beta_2 = \frac{v}{R_2} - \frac{1}{A_2}\frac{\partial w}{\partial \phi}. \tag{5.13b}$$

The quantities $A_1 = |\frac{\partial p_0}{\partial \alpha}|$ and $A_2 = |\frac{\partial p_0}{\partial \phi}|$ are the fundamental form parameters, R_1 and R_2 are the principal radii of curvature.

Contact Forces

The forces F_i originate from the frictional contacts and eventual elastic supports. In the simplest case, when they originate from massless friction pins, as shown in Fig. 5.1, they can be calculated from a force balance on the pin, as depicted in Fig. 5.2. In order to obtain the virtual work of the contact forces, the velocities of the contact points have to be derived. A point on the surface of the shell is parameterized through a point on the neutral surface

$$p_P(\alpha, \phi) = p_M(\alpha, \phi) + \frac{h}{2}e_\nabla(\alpha, \phi) \tag{5.14}$$

with

$$e_\nabla(\alpha, \phi) = \frac{\frac{\partial p_M}{\partial \alpha} \times \frac{\partial p_M}{\partial \phi}}{\|\frac{\partial p_M}{\partial \alpha} \times \frac{\partial p_M}{\partial \phi}\|}. \tag{5.15}$$

In order to calculate the virtual velocity of the material point on the shell currently in contact with the pad, the corresponding point on the neutral surface has to be determined, as can be seen in Fig. 5.3. The coordinates $\alpha_P = \alpha_0 + \Delta\alpha$, $\phi_P = \phi_0 + \Delta\phi$ parameterizing the contact in the deformed configuration, differ from the coordinates of the contact point in the undeformed configuration, α_0, ϕ_0 by $\Delta\phi$, $\Delta\alpha$. Note that depending on which coordinate system is used for the description of the problem, α_0 and ϕ_0 may or may not depend explicitly on time. From Fig. 5.3 we see that $\Delta\phi$, $\Delta\alpha$ are determined by the equations

$$\Delta\alpha = -\frac{h}{2}w_{,\alpha}(\alpha_0, \phi_0, t) + u(\alpha_0, \phi_0, t) + \mathcal{O}(u, w), \tag{5.16a}$$

$$\Delta\phi = -\frac{h}{2}w_{,\phi}(\alpha_0, \phi_0, t) + v(\alpha_0, \phi_0, t) + \mathcal{O}(v, w). \tag{5.16b}$$

Fig. 5.2 Force balance at the friction pin

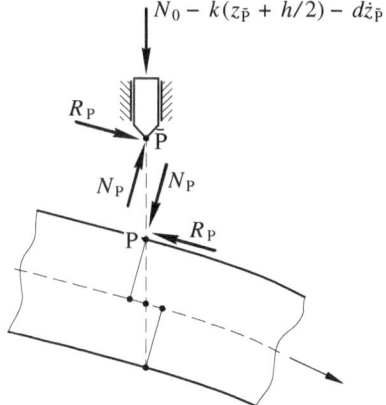

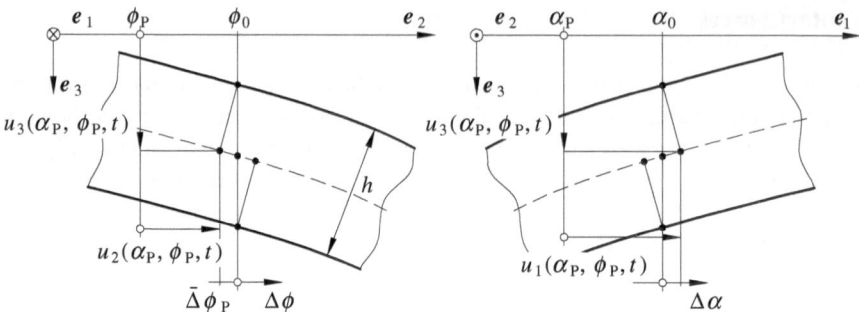

Fig. 5.3 Kinematics

Therefore, we have

$$w(\alpha_0 + \Delta\alpha, \phi_0 + \Delta\phi, t) = w(\alpha_0, \phi_0, t) + \mathcal{O}(u, v, w), \qquad (5.17)$$

which means that the quantities $\Delta\phi$, $\Delta\alpha$ do not enter the linearized equations.

5.1.2 Discretization

For the calculations in the following sections the functions u, v and w are expanded

$$u(\alpha, \phi, t) = \sum_{i=1}^{I_u} U_i(\alpha, \phi)q_i^u(t), \qquad (5.18a)$$

$$v(\alpha, \phi, t) = \sum_{i=1}^{I_v} V_i(\alpha, \phi)q_i^v(t), \qquad (5.18b)$$

$$w(\alpha, \phi, t) = \sum_{i=1}^{I_w} W_i(\alpha, \phi)q_i^w(t) \qquad (5.18c)$$

in shape functions satisfying at least the geometric boundary conditions. Linearization of the discretized equations with respect to the q_i and $\dot{q}_i$ in the general case yields equations of the form

$$M\ddot{q} + (G(t) + D(t))\dot{q} + (K(t) + N(t))q = f$$

where

$$M(t) = M^T(t) > 0, \qquad G(t) = -G^T(t), \qquad D(t) = D^T(t) \geq 0,$$
$$K(t) = K^T(t) > 0, \qquad N(t) = -N^T(t), \qquad f = \text{const.}$$

Depending on the coordinates used to describe the problem, the matrices usually periodically depend on time. In analogy to the investigations for the rigid rotor in Sect. 4.4, depending on the choice of problem representation, some of the matrices are constant. Using coordinates rotating with the body, for moderate revolution speeds the gyroscopic terms are small compared to the elastic restoring terms, so that the results of Chap. 2 remain applicable.

5.2 Avoidance of Self-Excited Vibrations in a Beam with Periodic Boundary Conditions

As a special case we now consider a circular beam on an elastic bedding, which is in contact with idealized friction pads. If we consider a rotating beam and spatially fixed pins as shown in Fig. 5.4, this is can be regarded as a minimal model for disk brake squeal [16]. On the other hand we can consider a spatially fixed beam in contact with pads moving in the opposite direction. This can serve as a minimal model for a singing wine glass. If the effect of centrifugal stiffening is neglected, it is easily seen that both problems can be described by the same equations of motion. At first we consider a rotationally symmetric beam. The equations of motion have a very simple form when they are set up with respect to a frame which is attached to the load. For the rotating beam this means that we set up the equations of motion with respect to a spatially fixed reference frame, for the rotating load we work in a coordinate system rotating with the load. In both cases the corresponding boundary value problem for a circular beam with radius a reads

$$\rho A \left[\ddot{w}(\phi, t) + 2 \frac{\Omega}{a^2} \dot{w}'(\phi, t) + \frac{\Omega^2}{a} w''(\phi, t) \right] + \frac{EI}{a^4} w^{IV}(\phi, t) + c w(\phi, t) = 0,$$

(5.19a)

$$w(0, t) - w(2\pi, t) = 0,$$

(5.19b)

Fig. 5.4 Rotating beam in frictional contact

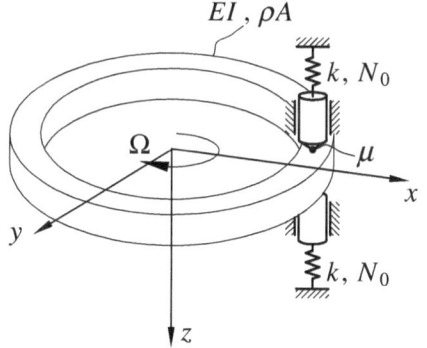

$$w'(0,t) - w'(2\pi, t) = 0, \tag{5.19c}$$

$$\frac{EI}{a^2}[w''(0,t) - w''(2\pi, t)] = M(w(0,t), w'(0,t)), \tag{5.19d}$$

$$\frac{EI}{a^3}[w'''(0,t) - w'''(2\pi, t)] = Q(w(0,t), w'(0,t)), \tag{5.19e}$$

where the contact forces $Q(w(0,t), w'(0,t))$ and torques $M(w(0,t), w'(0,t))$, originating form the contact forces at $\phi = 0$, can be derived as described in the previous section and are similar to the ones derived in [24,79]. For the current example they read

$$Q = -2kw(0,t), \tag{5.20a}$$

$$M = h\left(\frac{N_0}{a}\mu^2 w'(0,t) - k\mu w(0,t)\right). \tag{5.20b}$$

$A(w(0,t), w'(0,t))$ and $M(w(0,t), w'(0,t))$ are homogeneous functions in w and w' and can be linearized for a stability analysis. If the equations of motion are set up with respect to a frame attached to the beam, i.e. in material coordinates, in both cases the boundary value problem reads

$$\rho A\ddot{w}(\phi, t) + \frac{EI}{a^4}w^{IV}(\phi, t) + cw(\phi, t) = 0, \tag{5.21a}$$

$$w(0,t) - w(2\pi, t) = 0, \tag{5.21b}$$

$$w'(0,t) - w'(2\pi, t) = 0, \tag{5.21c}$$

$$\frac{EI}{a^2}[w''(\Omega t^-, t) - w''(\Omega t^+, t)] = M(w(\Omega t, t), w'(\Omega t, t)), \tag{5.21d}$$

$$\frac{EI}{a^3}[w'''(\Omega t^-, t) - w'''(\Omega t^+, t)] = Q(w(\Omega t, t), w'(\Omega t, t)) \tag{5.21e}$$

and not only has periodic boundary conditions in space but also with respect to time. As before we interpret the terms originating form the pads as perturbations. The discretized equations of motion for the statement of the problem in a frame attached to the load have the form

$$M\ddot{q} + G\dot{q} + (K + \Delta K)q = 0, \quad M = M^T, \ G = -G^T, \ K = K^T, \ \Delta K \neq \Delta K^T. \tag{5.22}$$

In a coordinate system attached to the beam the discretized equations read

$$M\ddot{q} + (K + \Delta K(t))q = 0, \quad M = M^T, \ K = K^T, \ \Delta K(t) \neq \Delta K^T(t) \tag{5.23}$$

and have precisely the form of the equations studied in Chap. 2. Therefore in order to avoid self-excited vibrations we want to design the beam such that no multiple eigenfrequencies in the unperturbed problem occur. An important question to ask is of course how much the eigenfrequencies have to be split in order to stabilize the system. The answer is given by the approximations to the stability boundary given in Chap. 2. If the terms originating from the pads are small compared to the elastic restoring torques and we assume the damping in the system to be proportional to the stiffness from (2.27) it follows

$$\Delta k = 2\sqrt{n^2 - d^2\omega^2}, \tag{5.24}$$

where $\Delta k = 2\pi \Delta \omega^2$, $d = 2\pi \alpha \omega^2$ and n can be estimated from the contact forces and the eigenforms of the unperturbed problem. In the present case n can be estimated as

$$n \le \frac{2\pi \frac{h}{a} k \mu \hat{w}_i \hat{w}'_j}{\int_0^{2\pi} \rho A w_i^2 a \, d\phi}, \tag{5.25}$$

where $\hat{w}'_i$, $\hat{w}_j$ are the maxima belonging to the slope and the displacement of the double modes of the unperturbed problem. From (5.24) it can furthermore be seen that in case of proportional damping the higher modes are more damped such that it is sufficient to obtain a frequency splitting for the lower modes. We now turn to constructive measures to obtain a frequency splitting. For the case of the rotating beam we want to find modifications which are balanced. On the other hand for a moving load problem a balance condition is not important. Therefore we start with this simpler case. A very simple modification, which is easy to realize in practice, is to attach a particle of mass m to the beam at the position $\phi = 0$. The corresponding boundary value problem without pads reads

$$\rho A \ddot{w} + \frac{EI}{a^4} w^{IV} + cw = 0, \tag{5.26a}$$

$$w(0, t) = w(2\pi, t), \qquad w'(0, t) = w'(2\pi, t), \tag{5.26b}$$

$$w''(0, t) = w''(2\pi, t), \qquad \frac{EI}{a^2}[w'''(0, t) - w'''(2\pi, t)] = m\ddot{w}(0, t). \tag{5.26c}$$

As described for example in [13], this elementary boundary value problem can be solved exactly using the ansatz

$$w(\phi, t) = W(\phi)e^{i\omega t}. \tag{5.27}$$

Substitution into (5.26a) yields the linear spatial differential equation

$$W^{IV} - \beta^4 W = 0 \tag{5.28}$$

with $\beta^4 = \frac{(\rho A \omega^2 - c)a^4}{EI}$. For $\beta \ne 0$ the general solution is

$$W(\phi) = C_1 \sin \beta\phi + C_2 \cos \beta\phi + C_3 \sinh \beta\phi + C_4 \cosh \beta\phi. \tag{5.29}$$

Adjusting this solution to the boundary conditions and requiring the existence of nontrivial solutions yields the characteristic equation

$$f(\beta, m) = 16\beta^3 \sin(\pi\beta) \sinh(\pi\beta) \left[\sin(\pi\beta) \cosh(\pi\beta) m \left(\tilde{c} + \tilde{d}\beta^4 \right) \right.$$
$$\left. + \sinh(\pi\beta) \left(\cos(\pi\beta) m \left(\tilde{c} + \tilde{d}\beta^4 \right) - 4\beta^3 \sin(\pi\beta) \right) \right] = 0 \quad (5.30)$$

with $\tilde{c} = \frac{c}{EI\rho A}$ and $\tilde{d} = \frac{1}{\rho A a^4}$. Note that for $m = 0$ we obtain the characteristic equation

$$64\beta^6 \sin^2(\pi\beta) \sinh^2(\pi\beta) = 0 \quad (5.31)$$

corresponding to the free circular beam. Equation (5.31) clearly has infinitely many double zeros $\beta_{i0} = k \in \mathbb{N}$ due to the quadratic sine term which yield double eigenfrequencies. The subscript 0 indicates that we are referring to the rotationally symmetric beam with $m = 0$. In Fig. 5.5 the characteristic equation is shown for the parameter $m = 0$ in blue and $m = 0.01$ in red for different scalings. In this case all other parameters have been chosen to unity. It can be seen that the double zeros of the characteristic equation split into single ones as expected. It can further be seen that

$$\frac{d}{d\beta} f(\beta, m)|_{\beta=\beta_{i0}} = -16\pi m \beta_{i0}^3 \cos^2(\pi\beta_{i0}) \sinh^2(\pi\beta_{i0}) \left(c + d\beta_{i0}^4 \right) \neq 0,$$
$$(5.32)$$

which means that for small masses m all eigenfrequencies are split. It is to be expected that this property also carries over to more complicated shell models. In fact, when investigating a singing wine glass in the experiment in Sect. 6.5.3 we make a similar observation.

Now we turn to the problem of the rotating beam for which we also want to split the spectrum. However, we additionally require that the modifications do not cause an unbalance. One way to achieve this goal is to modify the bending stiffness EI of the beam, which we want to do continuously. Therefore it is no longer possible to study the boundary value problem and we have to work with the discretized equations of motion. As shape functions we use the eigenfunctions corresponding to the free circular beam

$$w_k^c = \cos(k\phi), \qquad w_k^s = \sin(k\phi).$$

From the structure of eqs. (5.23) and the analysis in Chap. 2, it is clear that one should separate the pairs of semi-simple eigenvalues to avoid self-excited vibrations. In varying EI for mathematical convenience we agree to leave the area of the cross-section constant and to just vary the shape. Furthermore, we would like to keep our modifications point symmetric with respect to the axis of rotation, in order to keep the structure balanced. The discretized equations of the beam expanded in $w_k^c = \cos(k\phi)$ and $w_k^s = \sin(k\phi)$ read

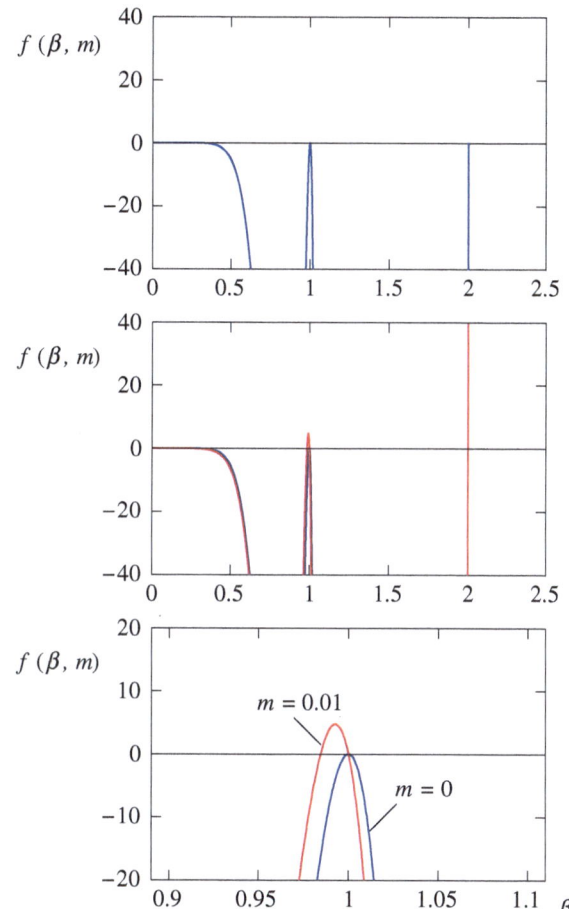

Fig. 5.5 Characteristic equation for $m = 0$: *blue* and $m = 0.01$: *red*

$$\begin{bmatrix} M^c & 0 \\ 0 & M^s \end{bmatrix} \ddot{q} + \begin{bmatrix} K^c & 0 \\ 0 & K^s \end{bmatrix} q = 0, \tag{5.33a}$$

with

$$m_{ij}^c = \rho A \pi, \qquad m_{ij}^s = \rho A \pi, \tag{5.33b}$$

$$k_{ij}^c = \int_0^{2\pi} \left(\frac{EI(\phi)}{a^4} w_i''^c w_j''^c + c w_i^c w_j^c \right) d\phi, \tag{5.33c}$$

$$k_{ij}^s = \int_0^{2\pi} \left(\frac{EI(\phi)}{a^4} w_i''^s w_j''^s + c w_i^s w_j^s \right) d\phi. \tag{5.33d}$$

A good solution for the choice of bending stiffness is given by

$$EI(\phi) = EI_0 + \sum_{k=1}^{N} EI_k \cos^2 k\phi = EI_0 + \sum_{k=1}^{N} EI_k \frac{1}{2}(1 + \cos 2k\phi), \qquad (5.34)$$

which is a point symmetric modification.

Making use of the orthogonality relations for trigonometric functions

$$\int_0^{2\pi} \cos(2k\phi)\cos(l\phi)\cos(m\phi)d\phi = \begin{cases} \dfrac{\pi}{2} & l+m = 2k \\ \dfrac{\pi}{2} & m-l = 2k \\ \dfrac{\pi}{2} & l-m = 2k \\ 0 & \text{otherwise} \end{cases}$$

$$\int_0^{2\pi} \cos(2k\phi)\sin(l\phi)\sin(m\phi)d\phi = \begin{cases} -\dfrac{\pi}{2} & l+m = 2k \\ \dfrac{\pi}{2} & m-l = 2k \\ \dfrac{\pi}{2} & l-m = 2k \\ 0 & \text{otherwise} \end{cases}$$

$$\int_0^{2\pi} \cos(2k\phi)\sin(l\phi)\cos(m\phi)d\phi = 0$$

we see that the terms $EI_k\frac{1}{2}(1 + \cos 2k\phi)$ influence the matrices corresponding to the sine and the cosine terms in Eq. (5.33a) differently on the main diagonal of the stiffness matrix.

Since K is diagonal dominant, the modification $EI_k \cos(2k\phi)$ yields

$$\omega_k^{*c2} \approx \omega_k^2 + \frac{EI}{2\rho h_k}\left(\frac{k}{a}\right)^4, \qquad \omega_k^{*s2} \approx \omega_k^2 - \frac{EI}{2\rho h_k}\left(\frac{k}{a}\right)^4,$$

which can also be derived arguing with perturbation theory and interpreting EI_k as a perturbation parameter. We see that the eigenfrequency of the cosine mode is increased while the eigenfrequency of the sine mode is decreased. Since the functions $\cos k\phi$ form a base of $L^2_{\text{sym}}(0, 2\pi)$, the functions $\cos 2k\pi$ form a base of all point symmetric functions on $L^2(0, 2\pi)$. So at least up to a constant, any point symmetric shape can be constructed by (5.34).

Although being efficient from a mathematical perspective, the modifications of the circular beam may not be easy to realize from a manufacturing point of view. Therefore the results obtained in this section are more of a theoretical value, showing that modifications exist which almost exclusively affect one pair of semi-simple eigenvalues, without influencing the other pairs.

5.3 Occurrence and Avoidance of Gyroscopic Terms in Continuous Systems

In this section we want to investigate under which circumstances gyroscopic terms can be eliminated in continuous systems. Gyroscopic terms in rotating continua depend on the formulation of the equations of motion. One source of gyroscopic terms is the modeling of transport of material, the other one is CORIOLIS effects.

5.3.1 Transport Effects

Gyroscopic terms due to transport effects arise for example in the rotating beam with periodic boundary conditions in Sect. 5.2, where equations of motion were set up with respect to a nonrotating frame. The position vector of a material point is given by

$$p = re_r + w(\phi, t)e_z, \tag{5.35}$$

where e_r, e_ϕ are nonrotating polar coordinates. As discussed in Sect. 5.1, when calculating the material velocity, the material coordinates have to remain fixed i.e.

$$\dot{p} = \frac{d}{dt} w(\eta + \Omega t, t)e_z = (w_{,t} + \Omega w_{,\phi})e_z. \tag{5.36}$$

When formulating the kinetic energy and expanding w in shape functions

$$w(\phi, t) = \sum_i W_i(\phi)q_i(t), \tag{5.37}$$

it is seen that the kinetic energy contains terms linear in $\dot{q}_i$, which yield gyroscopic terms. The gyroscopic terms originating from the material transport can be avoided by formulating the equations of motion in rotating coordinates meaning that $\phi = \eta$. However, as we shall see in the next section, in the general case gyroscopic terms still arise due to CORIOLIS forces.

5.3.2 Coriolis Forces

In the last section we saw that gyroscopic terms arise naturally in moving continua if the equations of motion of a system are not set up in a reference frame which is moving with the undeformed configuration. In order to avoid gyroscopic terms we therefore set up the equations of motion in material coordinates. Consider a continuous system which is rotating about an axis at constant angular velocity Ω.

Without loss of generality we can assume the axis of rotation to be the e_z axis of a cartesian coordinate system rotating with the body. The position vector of a material point in the continuum can be formulated as

$$p = (x + u)e_x + (y + v)e_y + (z + w)e_z, \qquad (5.38)$$

where x,y,z are the material coordinates and $u = u(x, y, z, t)$, $v = v(x, y, z, t)$ and $w = w(x, y, z, t)$. Taking into account that $\dot{e}_x = \Omega e_y$ and $\dot{e}_y = -\Omega e_x$, the velocity and acceleration can be calculated as

$$\dot{p} = [\dot{u} - \Omega(y + v)]e_x + [\dot{v} + \Omega(y + v)]e_y + \dot{w}e_z, \qquad (5.39a)$$

$$\ddot{p} = [\ddot{u} - 2\Omega\dot{v} - \Omega^2(x + u)]e_x + [\ddot{v} + 2\Omega\dot{u} - \Omega^2(y + v)]e_y + \ddot{w}e_z. \qquad (5.39b)$$

Expanding u, v and w in appropriate shape functions

$$u = \sum_i U_i q_i^u, \qquad (5.40a)$$

$$v = \sum_i V_i q_i^v, \qquad (5.40b)$$

$$w = \sum_i W_i q_i^w \qquad (5.40c)$$

and noting that the virtual velocities of the material points read

$$\frac{\partial p}{\partial q_i^u} = U_i, \qquad \frac{\partial p}{\partial q_i^v} = V_i, \qquad \frac{\partial p}{\partial q_i^w} = W_i \qquad (5.41)$$

the inertia terms of the continuum in the equations of motion read

$$\begin{bmatrix} \int \rho U_i U_j dV & 0 & 0 \\ 0 & \int \rho V_i V_j dV & 0 \\ 0 & 0 & \int \rho W_i W_j dV \end{bmatrix} \begin{bmatrix} \ddot{q}_j^u \\ \ddot{q}_j^v \\ \ddot{q}_j^w \end{bmatrix} +$$

$$2\Omega \begin{bmatrix} 0 & \int \rho U_i V_j dV & 0 \\ -\int \rho U_j V_i dV & 0 & 0 \\ 0 & 0 & 0 \end{bmatrix} \begin{bmatrix} \dot{q}_j^u \\ \dot{q}_j^v \\ \dot{q}_j^w \end{bmatrix} -$$

$$\Omega^2 \begin{bmatrix} \int \rho U_i U_j dV & 0 & 0 \\ 0 & \int \rho V_i V_j dV & 0 \\ 0 & 0 & \int \rho W_i W_j dV \end{bmatrix} \begin{bmatrix} q_j^u \\ q_j^v \\ q_j^w \end{bmatrix} - \Omega^2 \begin{bmatrix} \int \rho x U_j dV \\ \int \rho y V_j dV \\ 0 \end{bmatrix}.$$

$$(5.42)$$

Obviously the equations of motion feature gyroscopic terms which occur in the degrees of freedom perpendicular to the axis of rotation. The structure of the gyroscopic matrix in (5.42) strongly depends on the shape functions chosen. If the boundary conditions of the continuum are such that the same orthonormalized shape functions can be used for the expansion of u and v, then the gyroscopic matrix only features terms on its anti diagonal. Using the same transformation as in Sect. 4.5, the gyroscopic terms can be eliminated from the discretized equations of motion.

5.4 Summary

In this chapter we have considered continuous rotors in frictional contact. The examples studied showed that self-excited vibrations can be avoided, if the spectrum of the rotor does not have multiple eigenvalues. As in the discrete case, we have discussed the influence of the choice of generalized coordinates on the mathematical structure of the equations of motion. In particular the role of gyroscopic terms has been discussed.

Chapter 6
Structural Optimization of a Disk Brake

In the last chapters we have seen discrete and continuous examples for structures which can be stabilized by splitting multiple eigenfrequencies. We have seen a discrete model for a disk brake and a model for a shell with a sliding contact, which precisely had the structure of the equations investigated in Chap. 2. In this chapter we will first derive a continuous model of a disk brake. Afterwards we discuss methods and approaches to design a brake disk with a well split spectrum under the constraint that it has to be balanced as a rigid body. We first investigate structural modifications of the semi-analytic model of a rotating KIRCHHOFF plate and exploit the scope and the limitations. In a next step, we analyze a procedure based on a finite element approach with commercial software. Finally we propose a method to effectively parameterize structural modifications for the optimization procedure and report on experiments.

Since the results in Chap. 3 reveal a very complex mathematical structure of the optimization problems, the results are local optima achieved after variation of the initial guesses.

6.1 Disk Brake Model

In this section we consider an analytic model of a disk brake, which can be seen as a special case of the rotating shell investigated in Chap. 5. We consider the rotating continuum with KIRCHHOFF kinematics shown in Fig. 6.1, which, for a completely symmetric rotor, has been previously studied in [77], where it has been shown that in- and out-of-plane vibrations decouple. Therefore we concentrate on the out-of-plane vibrations, which we refer to as plate vibrations. The brake rotor is in contact with pointwise elastic friction pads consisting of massless pins which are pressed on the surface by prestressed springs. The discretized linear equations can be derived as for the rotating shell. In a coordinate system rotating with the undeformed plate they read

$$M\ddot{q} + D(t)\dot{q} + (K + N(t))q = 0, \tag{6.1}$$

G. Spelsberg-Korspeter, *Robust Structural Design against Self-Excited Vibrations*, SpringerBriefs in Applied Sciences and Technology, DOI: 10.1007/978-3-642-36552-2_6, © The Author(s) 2013

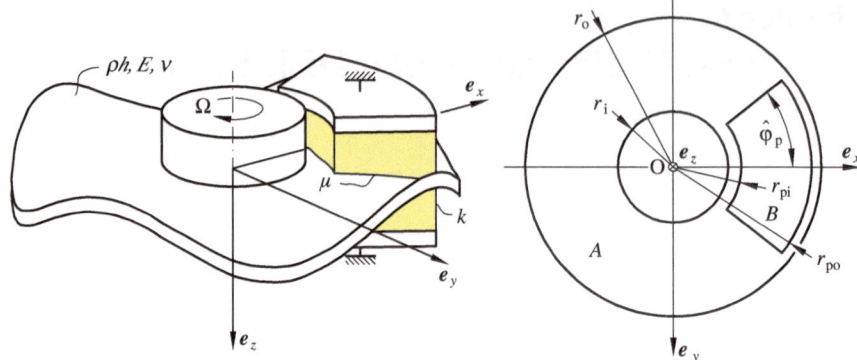

Fig. 6.1 Kirchhoff plate in distributed frictional contact

where $M = \mathrm{diag}(M_i)$, $K = \mathrm{diag}(\omega_i^2 M_i)$ and M_i is the normalization factor to the i-th eigenform corresponding to eigenfrequency ω_i of the nonrotating disk. The matrices $D(t) = D(t + 2\pi/\Omega)$ and $N(t) = N(t + 2\pi/\Omega)$ are periodic matrices and the plate is discretized by

$$w(r, \varphi, t) = \sum_{i=1}^{N} W_i(r, \varphi)\, q_i(t) \tag{6.2}$$

using adequate shape functions $W_i(r, \varphi)$. The generic element of the matrices N and D are

$$
\begin{aligned}
N_{ij} = \int_B \Bigg[& 2k\, W^i W^j - \frac{hk\mu}{r} W^i_{,\varphi} W^j + (1 + \mu^2) \frac{hN_0}{r^2} W^i_{,\varphi} W^j_{,\varphi} - \frac{h^2 N_0 \mu}{2r^3} W^i_{,\varphi\varphi} W^j_{,\varphi} \\
& + \frac{h^2 N_0 \mu}{2r^2} \left(W^i_{,\varphi} W^j_{,r} - W^i_{,r} W^j_{,\varphi} \right) + hN_0 W^i_{,r} W^j_{,r} - \frac{h^2 N_0 \mu}{2r} W^i_{,\varphi r} W^j_{,r} \\
& + \frac{d\Omega h \mu}{r} W^i_{,\varphi} W^j_{,\varphi} - 2d\Omega W^i W^j_{,\varphi} \Bigg]\, dB
\end{aligned}
\tag{6.3}
$$

and

$$D_{ij} = \int_B \left[2d\, W^i W^j - \frac{dh\mu}{r} W^i_{,\varphi} W^j + \frac{h^2 N_0 \mu}{2r\Omega} W^i_{,r} W^j_{,r} \right] dB \tag{6.4}$$

respectively, where

$$\int_B (*)\, dB = \int_{r_{pi}}^{r_{pa}} \int_{-\hat{\varphi}_p - \Omega t}^{\hat{\varphi}_p - \Omega t} (*)\, r\, dr\, d\varphi, \tag{6.5}$$

The Eq. (6.1) have precisely the structure of the equations considered in Chap. 2. Therefore, in order to construct a disk which is robust against squeal, the eigenfrequencies should be split as far as possible. This will be done in the following.

The Eq. (6.1) can be also used for quantitative studies on asymmetric disks. A promising approach is to perform a finite element analysis on an asymmetric disk and to study the rotating disk with pads using the shape functions in (6.1). Due to the small rotational frequencies of the disk compared to the eigenfrequencies, the integration times required for FLOQUET analysis are however long and may yield numerical uncertainties. Nevertheless this method has been successfully used in [78, 90] to study asymmetries.

6.2 Semi-Analytic Approach

We consider the continuous model of the disk brake discussed in Sect. 6.1. The idea of the underlying optimization approach is that for a plate with grooves the discretization in global shape functions can be maintained and in order to model the structural modifications only the energy expressions have to be corrected.

6.2.1 Formulation of the Optimization Problem

The disk is discretized with a RITZ approach using the eigenfunctions of the nonrotating plate corresponding to eigenfrequencies in the relevant frequency range. Instead of evaluating the occurring integrals over the whole domain in one step, the plate is divided into sectors as demonstrated in Fig. 6.2. For each sector the height of the cooling rib enters the equations as a factor multiplying the occurring integrals.The discretized eigenvalue problem corresponding to the linearized equations of motion can therefore be written in a relatively simple form as

$$\det\left(\sum_n M_n(1-a_n)\lambda^2 + \sum_n K_n(1-a_n^3)\right) = 0, \tag{6.6}$$

$$(m_{ij})_n = \frac{\partial^2}{\partial \dot{q}_i \partial \dot{q}_j} \int_{-\frac{h}{2}}^{\frac{h}{2}} \int_{\varphi_{n-1}}^{\varphi_n} \int_{r_i}^{r_o} \rho \dot{w}(r,\varphi,t)^2 r \, dr \, d\varphi \, dz, \tag{6.7}$$

$$(k_{ij})_n = \frac{\partial^2}{\partial q_i \partial q_j} \int_{-\frac{h}{2}}^{\frac{h}{2}} \int_{\varphi_{n-1}}^{\varphi_n} \int_{r_i}^{r_o} u_{el}(r,\varphi,t) r \, dr \, d\varphi \, dz, \tag{6.8}$$

in which the heights of the grooves appear as variables. We have gained an approximation for the eigenvalue problem depending on the heights of the ribs. At this point we remark that, from a physical point of view, the formulation only makes sense

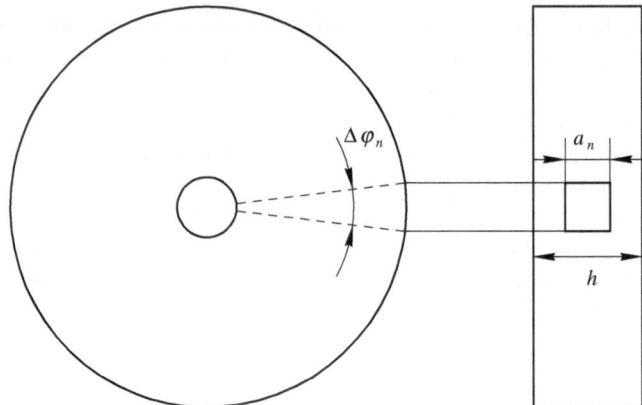

Fig. 6.2 Division of the brake disk in sectors

if the heights of the cross-sections are chosen in such a way, that the KIRCHHOFF hypothesis for the cross-sections stays approximately valid.

Based on the discretized eigenvalue problem (6.6), we can formulate an optimization problem

$$\max_{a_n} \tau \tag{6.9a}$$

$$\text{s.t.} \tag{6.9b}$$

$$\frac{\omega_{k+1} - \omega_k}{\omega_k} \geq \tau \qquad \omega_{\hat{N}} \geq \omega_{\hat{N}-1} \geq \cdots \geq \omega_1 \tag{6.9c}$$

$$0 \leq a_n \leq \hat{a} \text{ for all } n, \tag{6.9d}$$

$$a_n = 0 \quad \text{for selected } n, \tag{6.9e}$$

$$\sum_n a_n \cos \phi_n = 0, \tag{6.9f}$$

$$\sum_n a_n \sin \phi_n = 0 \qquad \phi_n = \left(n - \frac{2\pi}{N}\right). \tag{6.9g}$$

The objective function is chosen to maximize the minimum relative distance between the eigenfrequencies of the rotor. The constraints ensure that the disk stays balanced, the height of the channel is less than $\hat{a} \leq h$ and that at certain places no channels occur.

The optimization problem is solved with the MATLAB optimization toolbox using the parameter values (non-dimensionalized with the dimensionless time $\bar{t} = \frac{t}{r_o}\sqrt{\frac{E}{\rho}}$ and the radius $\bar{r} = \frac{r}{r_o}$)

$$\bar{r}_i = \frac{r_i}{r_o} = 0.154, \qquad \bar{r}_o = \frac{r_o}{r_o} = 1, \qquad \bar{h} = \frac{h}{r_o} = 0.1605, \qquad \nu = 0.3, \tag{6.10}$$

Fig. 6.3 Optimized disk (for $N = 36$, $\hat{a} \leq 0.8\,h$, min $\frac{\omega_{k+1}-\omega_k}{\omega_k} \approx 3.1\,\%$, 6 channels placed symmetric over the circumference are kept solid by constraint)

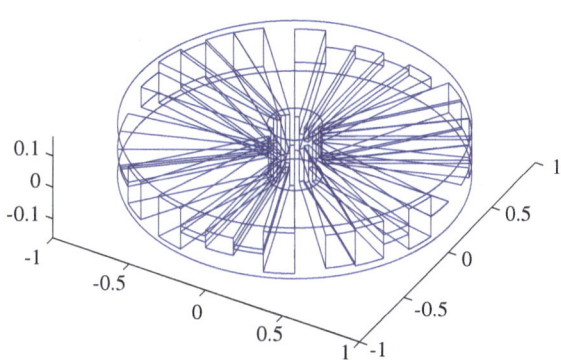

which are based on parameters for a disk brake used in [24]. They correspond to a brake rotor represented by an equivalent KIRCHHOFF plate with $r_o = 0.162$ m, $E = 4.16 \cdot 10^{10}$ N/m^2 and $\rho = 4846$ kg/m^3. These parameters were experimentally identified in [32].

Figure 6.3 shows an optimized disk using 39 shape functions corresponding to eigenfrequencies in the range up to 17 kHz of the solid plate, and 36 sectors. This covers the audible frequency range. In this configuration it was possible to obtain a minimum relative difference for the optimized eigenfrequencies of 3.1 percent. By limiting the optimization to sectors of a plate, the design freedom is only exploited in a very narrow manner. In [93] the design space has been enlarged by introducing two sectors on the radius which can be optimized independently. Furthermore, different optimization strategies have been exploited including analytic gradients of the eigenvalues which can be calculated from the perturbation theory of eigenvalues [68].

6.2.2 Discussion on the Applicability of Plate Theory in the Context of Structural Modifications

The purpose of this section is to investigate in what range KIRCHHOFF theory remains applicable if grooves are allowed, as in the optimization problem in the last section. Therefore the plate with parameters from the previous section is investigated with one groove placed over the diameter as shown in Fig. 6.4. The groove is either placed outside or inside the disk and covers an angle of 10°. For both settings the eigenfrequencies and eigenmodes are calculated from KIRCHHOFF plate theory using the RITZ discretization from the previous section. The same plate is investigated with a standard finite Element approach using 3 D solid elements in ANSYS. In both cases the plate is assumed to be clamped on the inner and free on the outer radius. The height of the groove is varied form 0–20 % of the height of the disk. All of the

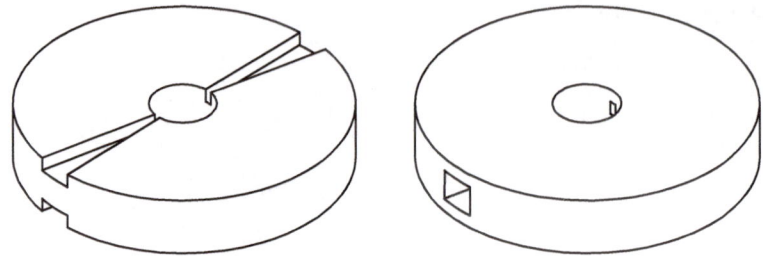

Fig. 6.4 Plate with inner and outer groove

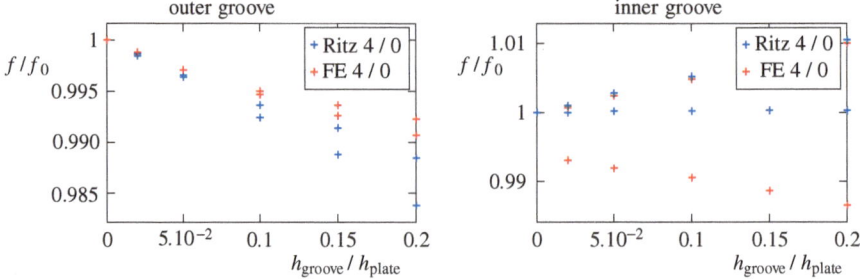

Fig. 6.5 Comparison of 3D elastic solution with Kirchhoff theory for a plate with an outer and an inner groove

FE calculations have been checked for convergence against the mesh size. It can be seen that the shape of the eigenforms of the plate is influenced relatively little by the modification. For the eigenform with four nodal diameters, the frequencies obtained from both methods are shown in Fig. 6.5 on the left. It is seen that at least for small heights of the modifications KIRCHHOFF theory and the proposed RITZ approach yields acceptable results. The same comparison for the inner groove is shown in Fig. 6.5 on the right. It is seen that even a modification of very small height yields significant differences between 3D elastic and KIRCHHOFF analysis. The reason for the qualitative difference between the comparison for the inner and the outer groove lies in the fact that it is impossible to enforce that planar crossections remain planar if no material lies on the neutral surface.

6.3 Design Space in a Finite Element Environment

The results of the previous section show that, when dealing with structural modifications, the scope of design parameters in which analytical models accurately describe their influence is quite limited. Therefore, in this section we investigate structural modifications on a brake rotor in a finite element environment. Again our goal is to split the spectrum as far as possible in the frequency range for brake squeal.

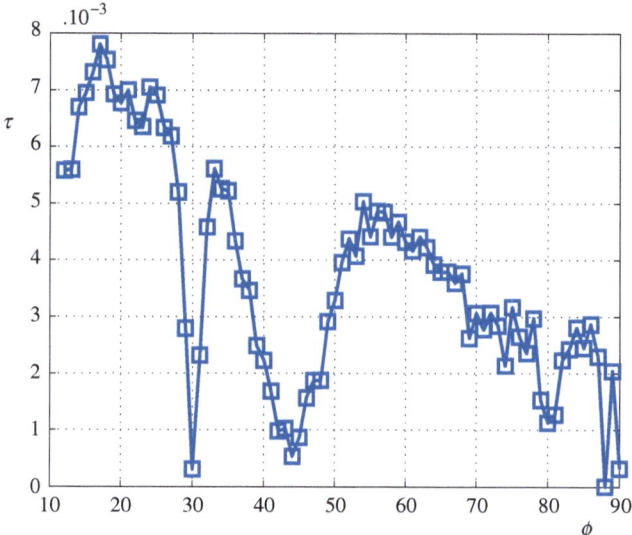

Fig. 6.6 Variation of the angle between two channels

Mathematically the optimization problem is formulated as

$$\max \tau, \tag{6.11a}$$

$$\tau = \min_k \frac{\omega_{k+1} - \omega_k}{\omega_k} \tag{6.11b}$$

for all ω_k in the relevant frequency range. In a first step, a cooling channel with quadratic cross section of 12.4 mm side length was placed on the diameter of $\phi = 0$. The cross section was chosen such that it corresponds to a circular one with diameter 14 mm in order to simplify the meshing process. In a second step, a second cooling channel was placed on the diameter at a variable angle ϕ which was varied from 0 to 90°. From Fig. 6.6 it can be seen that a minimum relative difference of the eigenfrequencies of 0.78 % could be achieved. It can also be seen that the minimum of the objective function has many local optima, which makes the optimization process difficult.

If an additional channel is placed at the angle ϕ_2 a variation of ϕ_1 and ϕ_2 yields the results presented in Fig. 6.7. The minimum relative difference of the eigenfrequencies can be increased to 0.87 %. Again, the complicated structure of the objective function can be seen. The best result was obtained for the placement of four cooling channels where one channel is kept fixed at $\phi_1 = 15°$ and the angles of two further channels ϕ_2 and ϕ_3 relative to the first channel at $\phi = 0$ were varied. The results for the placement of 4 channels are shown in Fig. 6.8. The minimum relative difference between two eigenfrequencies was increased to 1.3 % for the optimal configuration. Although the separation of eigenfrequencies achieved is quite low, the results are promising,

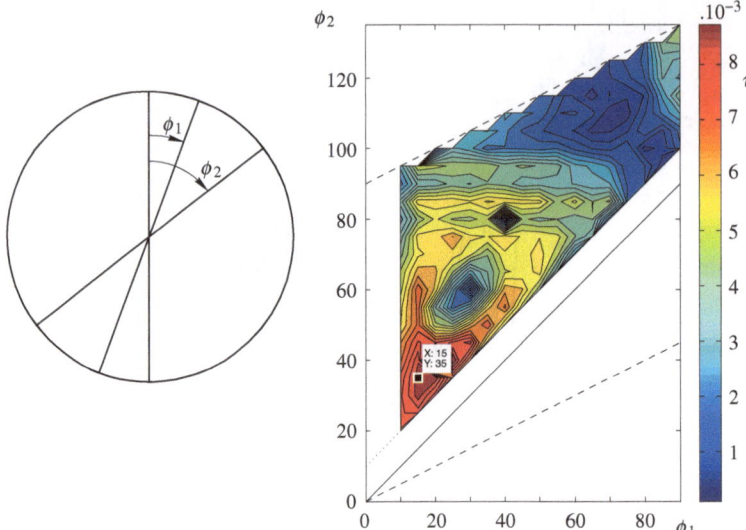

Fig. 6.7 Optimal placement of three (channels)

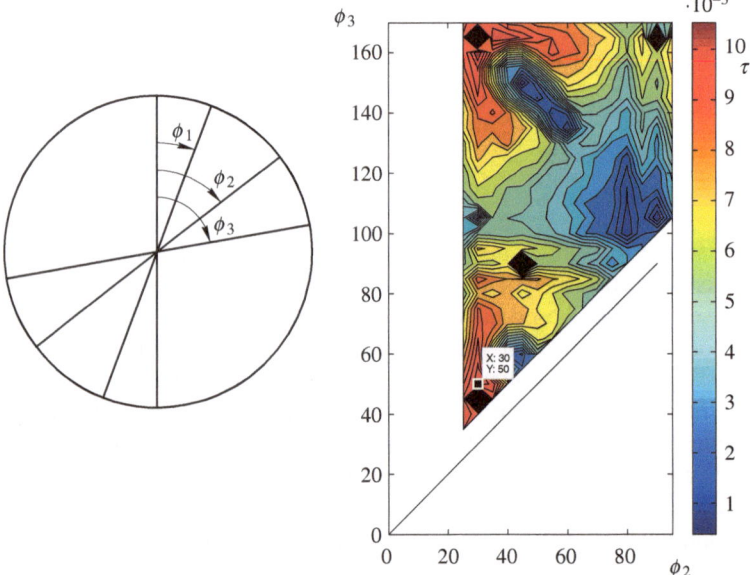

Fig. 6.8 Optimized placement of four channels

since the calculations on the wobbling disk model of Fig. 4.4 show that even small splitting of the eigenfrequencies significantly increases the stability region in the parameter space. At the same time it must be said that a brute force approach as the one mentioned is impractical because of the enormous computation times. The most problematic point is that the model has to be remeshed for every change in geometry. Even if instead of a simple variation of the angles an optimization procedure was to be followed, this is too costly. Therefore we will propose a new approach in the following.

6.4 Negative Stiffness Approach

The previous section showed that a standard optimization procedure for a disk brake with cooling channels is intractable if the modified geometry has to be remeshed in every step of the calculation. However, in many cases one has to deal with an easy underlying structure which is to be modified such that a certain modal behavior is achieved. The question therefore arises on how to describe the modifications in a convenient manner. Recently ILANKO introduced the concept of negative stiffness and mass in the context of asymptotic modeling [28, 29, 30, 31, 94]. We will adopt the idea of introducing negative mass and stiffness in order to model modifications of the structure. The advantage of the approach is that the underlying structure and the modifications can be discretized separately and the energy expressions of the underlying structure can be corrected by the contribution of the modifications. This means a simple coupling of two discretization schemes. For global shape functions a similar approach has been followed in [38].

6.4.1 Demonstration of the Approach with a Simple Example

The basic idea of the approach is that by using the concept of negative stiffness it is possible to discretize different parts of a structure in an adequate manner in a first step and to couple the discretizations in a second step.

As probably the easiest example consider the structural model of a longitudinal bar with different material parameters between L_1 and L_2 as shown in Fig. 6.9. The goal is to use one discretization for the global rod and one for the modification with different material parameters. The displacement field can be expanded in terms of

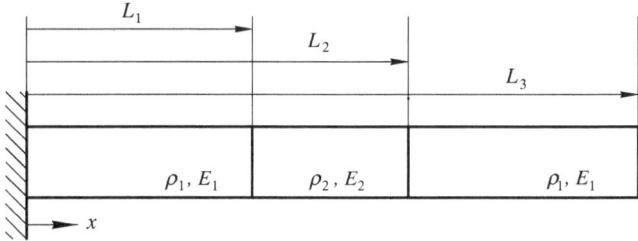

Fig. 6.9 Longitudinal bar with different sections

shape functions

$$u(x,t) = \sum_i W_i(x) q_i(t)$$

yielding the discretized equations of motion

$$M\ddot{q} + Kq = 0, \qquad m_{ij} = \int_0^{L_3} \rho A W_i W_j \mathrm{d}x, \qquad k_{ij} = \int_0^{L_3} E A W_i' W_j' \mathrm{d}x.$$

As shape functions one can either use global shape functions as for example $W_i(x) = \sin(i\frac{\pi}{L}x)$ or local shape functions

$$\begin{aligned} W_i(x) &= 1 + \frac{x - ih}{h} & (i-1)h \leq x \leq ih \\ &= -\frac{x - (i+1)h}{h} & ih \leq x \leq (i+1)h, \end{aligned}$$

where $h = \frac{L_3}{N}$ is the length of a local element. The domain between L_1 and L_2 is taken into account using a separate discretization scheme and using negative stiffness and mass with the stiffness matrix K^- and the mass matrix M^-.

In order to couple the two discretization schemes, the degrees of freedom of both schemes have to be coupled by kinematic constraints. Assuming that the modification between L_1 and L_2 is modeled with local shape functions, the degrees of freedom of the nodes $\tilde{q}_i$ at the positions $\tilde{x}_i$ are constrained by

$$\tilde{q}_i = w(\tilde{x}_i, t)$$

yielding

$$\tilde{q} = Qq.$$

This has to be taken into account in the corresponding energy expressions

$$T = \frac{1}{2}\dot{q}^T M \dot{q} - \frac{1}{2}\dot{\tilde{q}}^T M^- \dot{\tilde{q}},$$

$$U = \frac{1}{2}q^T Kq - \frac{1}{2}\tilde{q}^T K^- \tilde{q}.$$

The eigenvalue problem thus reads

$$[(M - Q^T M^- Q)\lambda^2 + (K - Q^T K^- Q)]v = 0.$$

Using the parameters

$\rho_1 = 7850 \text{ kg/m}^3,$ $\quad \rho_2 = 7850 \text{ kg/m}^3,$ $\quad E_1 = 2.06 \, 10^{11} \text{ N/m}^2,$ $\quad E_2 = 1.03 \, 10^{11} \text{ N/m}^2,$

$L_3 = 1 \text{ m},$ $\quad L_1 = 0.4 \text{ m},$ $\quad L_2 = 0.65 \text{ m},$

Table 6.1 Eigenfrequencies for different solution approaches in rad/s

	Global shape functions	Local shape functions
ω_1	7274	7229
ω_2	21625	21533
ω_3	37188	36921
ω_4	51544	51286
ω_5	65326	64753

Table 6.2 Nonzero eigenfrequencies for different solution approaches in rad/s

	Exact solution	Global shape functions	Local shape functions
ω_1	20117	6487	19938
ω_2	45981	24033	45535
ω_3	60350	33043	59863
ω_4	91962	52321	91208
ω_5	100580	70399	99927

where $E_2 = 0.5\,E_1$ we perform three numerical experiments. In one calculation we use 12 global shape functions for the discretization of the rod and for the second one we use 80 local ones. The area of the modification is approximated with 40 local shape functions. The results in Table 6.1 show that all results are in good agreement. Performing the same calculation using $E_2 = 0$, which physically means that the right part of the rod can move freely, reveals a very different result shown in Table 6.2. It is seen that the convergence of the approach with global shape functions is rather poor. The convergence for the approach with local shape functions is acceptable, however we note that the approximations do not yield an upper bound for the eigenfrequencies as for a standard RITZ-GALERKIN approach. A couple of approximately zero eigenvalues corresponding to the nodes in the modification have been discarded in the table. The same holds for a zero eigenvalue which accounts for the fact that the right part of the body can perform unrestrained rigid body motions. The reason for the poor convergence with global shape functions is that now the left and the right part of the rod are not connected, whereas the shape functions couple both parts. A look at the approximation of the first eigenform shows that it is very poorly approximated with the global shape functions. Nevertheless, in many applications the structure is not completely separated as in this case, and a good convergence is to be expected also with global shape functions.

6.4.2 Structural Optimization of a Disk Brake

The approach has been used in [80] to optimize a bicycle disk brake such that no double eigenfrequency occurs, by placing holes of given radius at variable positions. The disk shown in Fig. 6.10 has the inner radius $r_i = 67\,\text{mm}$ and outer radius

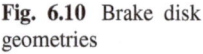

Fig. 6.10 Brake disk geometries

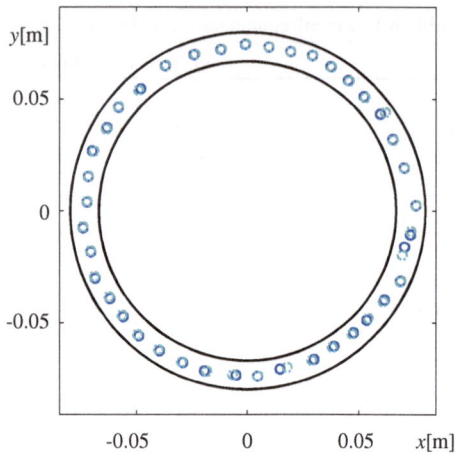

$r_a = 80\,\text{mm}$ and is $h = 2\,\text{mm}$ thick. We give a random initial placement (green circles) for 40 holes with a diameter of 4 mm and optimize for their positions (blue circles). In agreement with the discussion following formula (5.24) in Sect. 5.2 the objective function is

$$\max \min_{k}(\omega_{k+1} - \omega_k) \qquad \omega_{\hat{N}} \geq \omega_{\hat{N}-1} \geq \cdots \geq \omega_1, \qquad (6.12)$$

where the maximum frequency under consideration $\omega_{\hat{N}}$ is less than 6 kHz. The minimum distance between eigenfrequencies is maximized under the constraints that the disc stays statically and dynamically balanced and holes do not intersect with each other or the boundary. The corresponding geometry is shown in Fig. 6.10 and a minimum splitting of the eigenfrequencies of around 6 Hz has been achieved. The major advantage of the method is that for plates and holes the same discretization can be maintained throughout the optimization and that only the matrix Q has to be build up in every step of the optimization, which can be done efficiently.

In current industrial projects the method is also used to systematically optimize disk brakes for passenger cars. However the details of their development are part of a separate project and therefore omitted here.

6.5 Experimental Evidence

The idea of breaking symmetries in order to prevent noise is not new to design engineers. In this section we refer to experimental evidence which can be found in literature and to observations made at the brake test rig in Darmstadt. The experiments made are only preliminary in the sense that a systematic test for squeal under different external conditions is infeasible to perform at the test rig. However, a simple proof of concept can be achieved by an experiment with the singing wine glass. The introduction of a slight additional mass, for example by attaching magnets, destroys the excitation mechanism for the generation of sound.

6.5.1 Literature Review

In an early study on brake squeal MILLNER and NORTH suggested there should be a relation between disk brake squeal and a multiple mode of the brake rotor [47, 55, 56]. Based on this observation NISHIWAKI and coauthors designed disks with the goal to split eigenfrequencies of the squeal modes [54]. Analytically they investigated the placement of additional masses and in a finite element analysis they varied number and thickness of the cooling fins as well as the thickness of the rubbing plate. In experiments they used a disc in which two of 36 vanes were removed and another one, where two fins were eliminated. It was shown that brake squeal could be reduced significantly, however there were critical modes remaining in the relevant frequency range.

Around the same time, LANG and NEWCOMB experimented with structural modifications on drum brakes [12]. They also claim that with structural modifications an improvement on the squeal behavior could be achieved.

In a more recent work [9, 10] FIELDHOUSE and coworkers reconsidered the problem of finding appropriate modifications in order to prevent squeal. As structural modifications they considered the addition of masses and the drilling of holes. A drilled asymmetric disk was tested on a vehicle in operation for about 20 hours without showing squeal. Finally a cast iron design was developed which, apart from asymmetry, also took into account airflow and heat conduction. However no experimental results were reported on that particular disk.

Despite of the success in the experiments, in most of the designs used in [9, 10, 42, 54] there are still cyclic symmetries of an angle less than π present. Therefore at least from a theoretical point of view double eigenfrequencies are to be expected as shown in Sect. 3.3. However, these multiple eigenfrequencies are to be expected at high frequencies, which are out of the relevant range for squeal due to the dominance of damping in agreement with the reasoning in formula (5.24) in Sect. 5.2.

In a recent paper [46], MASSI and AKAY attached additional masses to a simplified disk brake test rig and showed that the distortion helped to prevent squeal. The phenomenon can intuitively be explained by the fact that the pad does not see the same stiffness at all times. In this work and the previous papers [72, 73] the underlying mechanism is explained more rigorously from a theoretical point of view also allowing for a quantitative analysis.

Fig. 6.11 Original and modified brake disk

6.5.2 Investigations on a Brake Test Rig

In order to get an experimental indication that the breaking of symmetries of the brake rotor helps to prevent squeal, a few preliminary experiments were conducted at the brake test rig in Darmstadt [75]. Since commercial brake disks are not completely symmetric due to the cooling channels, for the experiments they were not adequate to isolate the effect. Furthermore, the brake manufacturers use alloys for the cast iron which feature a relatively high material damping. This is one of the reasons which makes squeal difficult to reproduce in the experiment. In order to isolate the effect of breaking the symmetries of the brake rotor as much as possible two solid brake disks made of regular steel were manufactured and tested. The geometry was chosen identical to a commercial disk brake with the only difference, that no cooling channels were present. Using FE analysis structural modifications have been chosen with the goal of splitting as many eigenfrequencies as possible. It is observable that the splitting of all eigenfrequencies by a significant amount is not an easy task, even when considering only the audible frequency range. Relying on heuristic FE calculations, a modification with five holes through the whole disk has been chosen, as seen in Fig. 6.11. Note, that in accordance with the observations in Sect. 3.3 the design certifies that the disk does not have a cyclic symmetry of an angle less than π.

With a scanning laser vibrometer the spectrum of the plain and the modified disk mounted on the test rig without pads in a nonrotating configuration was measured using hammer excitation. A comparison of the spectra seen in Fig. 6.12 shows the splitting of the lowest multiple eigenfrequencies. After measuring the eigenfrequencies, the pads where applied to both disks in order to find a squealing configuration. During the experiment, an interesting observation was made: whereas the symmetric disk started to squeal almost immediately, squeal could not be produced with the non-symmetric disk, even after several attempts. In Fig. 6.13 one sees the two squealing frequencies measured in the experiment at approximately 4.2 and 7.5 kHz which are close to peaks of the disk without modifications. Furthermore it was observed that damping in the solid disks was much lower than in conventional disks. This reflects the effort that manufacturers have made to increase damping by choosing materials accordingly.

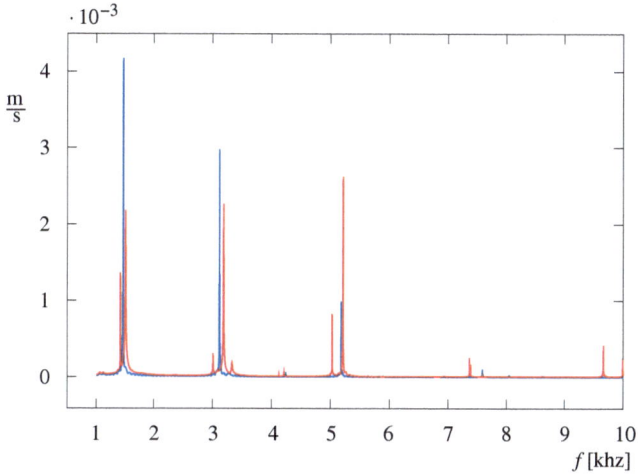

Fig. 6.12 Spectrum of the plain (*blue*) and modified (*red*) brake disk

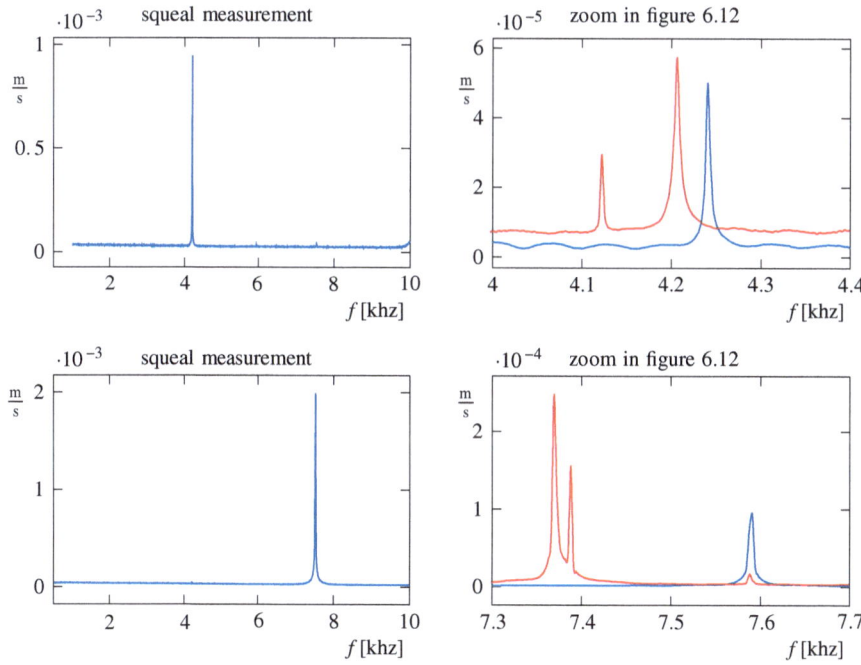

Fig. 6.13 Squealing mode and spectrum of the plain (*blue*) and modified (*red*) disk

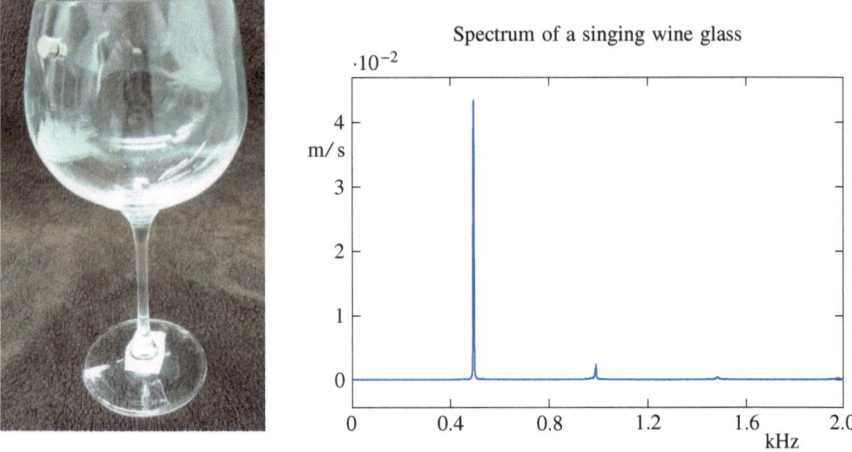

Fig. 6.14 Wine glass and spectrum

6.5.3 Investigation of a Singing Wine Glass

In this section we describe an experiment with a singing wine glass which is easily
reproducible in contrast to many experiments on the brake test rig. We took the
standard wine glass shown in Fig. 6.14 and wiped a finger on its surface to generate a
sound. The sound could be generated even easier when a wet rubber glove was used.
During the experiments the glass was filled with water up to a height of approximately
20 mm. For the experiment itself this is not essential, however it made the lubrication
of the rubber easier. We measured the corresponding response spectrum using a
scanning laser vibrometer. It is seen that primarily a mode at approximately 0.495 kHz
was excited and a few higher harmonics.

The experiment was repeated after two circular magnets (height: 4 mm, diameter:
8 mm) were attached to the glass as shown in Fig. 6.14. With the magnets attached
it was not possible to obtain a monofrequent sound and there was hardly any noise
generated when wiping the finger over the glass.

We measured the spectrum of the glass with and without magnets using hammer excitation. The corresponding graphs are shown in Fig. 6.15, where the blue
curve represents the plain wine glass and the red curve represents the wine glass
with magnets attached. In the red curve we find two peaks at each resonance over
0.1 kHz, whereas the blue curve only shows one per resonance. This indicates that
the eigenfrequencies have been significantly split and is in good agreement with our
theoretical predictions in Chap. 2 and Sect. 5.2. We note that the first peak is not split.
It occurs at a frequency below the audible range and probably is a bending mode of
the shaft of the wine glass, which is hardly affected by the magnets.

In order to emphasize the equivalence of the singing wine glass with a disk brake
we cut the shaft of a glass and attached it to a cordless screwdriver as shown in

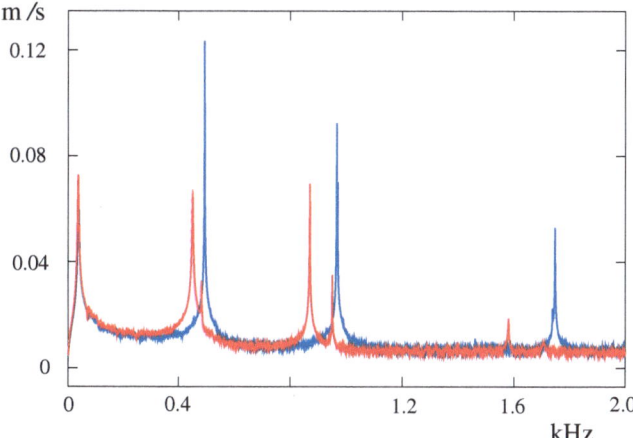

Fig. 6.15 Modal analysis of a wine glass with and without magnets

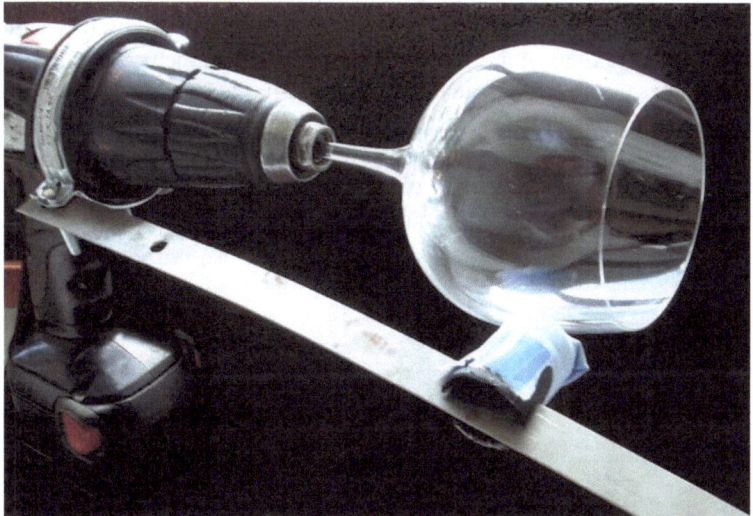

Fig. 6.16 Wineglass in cordless screwdriver

Fig. 6.16. Through an elastic lever a cartridge with a finger of a rubber glove is pressed onto the glass. When the rubber is lubricated with water, sound occurs for low operation speeds of the cordless screwdriver. When attaching the magnets again the system is quiet. We remark that for the occurrence of audible self-excited vibrations in the wine glass it is not necessary to wipe the glass on the rim. Other positions on the surface of the glass yield the same result. Summarizing, we observe that the relation between rotor asymmetry and squeal can be easily verified experimentally with a wine glass. In contrast to a commercial disk brake these experiments are far easier to conduct, since the phenomenon is more robust. This is of course due to the fact that commercial disk brakes are designed in order not to squeal and have relatively high damping.

6.6 Summary

In this chapter approaches for the structural optimization of a brake disk have been proposed. Apart from analytical approaches, which are very limited in the design space and in terms of their validity, approaches towards an optimization with finite elements and commercial software have been established. The concept of working with negative stiffnesses allows to work with different discretizations for an under-lying structure and modifications. The major advantage of the proposed approach is that a robust design against squeal is achieved without having to consider the precise formulation of the contact forces in a model. First of all they are intractable to model exactly, due to their dependence on several parameters and environmental changes, second even with a rough approximation such as COULOMB's law of friction, the time dependent equations of motion are hard to solve for larger systems. The theory developed shows that only the structure of the rotor has to be considered for the opti-mization, provided a sufficiently large splitting of the frequencies can be achieved. The method has been successfully applied to optimize disk brakes in an industrial project. First experiments verify the usefulness on the test rig. Of course we have to keep in mind that we have concentrated on the squeal problem in this book. In industry there are many other design requirements that are not yet considered in the present models. One example are thermal effects which will be investigated in future.

Chapter 7
Nonlinear Analysis of Systems Under Periodic Parametric Excitation

In Chap. 2 we investigated the stability behavior of linearized mechanical systems of the form

$$M\ddot{q} + \Delta D(t, \varepsilon)\dot{q} + (K + \Delta K(t, \varepsilon))q = 0. \tag{7.1}$$

We saw that the trivial solution could be stabilized by splitting the eigenfrequencies of the unperturbed problem (i.e. $\Delta D = 0$, $\Delta K = 0$). In Chap. 6 we used the idea to find a robust design for disk brakes against self-excited vibrations. All of the previous reasoning was however based on the study of the linearized equations of motion. It is clear that the usefulness of the approach does not only depend on the magnitudes of the perturbation matrices but also on the question for what magnitudes of the generalized coordinates q the linear equations give a good approximation to the nonlinear system. This can only be judged from the nonlinear behavior. Apart from the question of the scope of the linear model, it is interesting to study what happens in case of an instability of the linearized system. For systems with constant coefficients this is often studied using center manifold theory [85]. The case of time periodic equations of motion has had far less attention in the literature than the autonomous case. Approaches designed for the time-periodic case can be found in the papers of SINHA and coworkers (see for example [6, 95]). In the next section we will augment systems of the type (7.1) with nonlinearities. We will see that a center manifold exists and determines the long-term behavior of the system as in the constant coefficient case. A detailed study of the bifurcations will then be conducted through the expansion of the POINCARE map of the system. We will extend results of [22] and [23] for the qualitative bifurcation behavior of autonomous systems to systems of the type (7.1) augmented by nonlinearities. Parts of the results are also published in [76].

G. Spelsberg-Korspeter, *Robust Structural Design against Self-Excited Vibrations*, SpringerBriefs in Applied Sciences and Technology, DOI: 10.1007/978-3-642-36552-2_7, © The Author(s) 2013

7.1 Time Continuous Center Manifold Theory for Time-Periodic Systems

The equations of motion for systems of the type (7.1) augmented by time periodic nonlinearities can be written as

$$\dot{q} = A(t)q + f(t, q), \quad A(t) = A(t + T), \quad f(t, q) = f(t + T, q), \quad f(t, 0) = 0.$$
(7.2)

It is well known that the linear part of Eq. (7.2) can be transformed to constant JORDAN normal form [41], which is written here in real form for a system with two mechanical degrees of freedom

$$\dot{x} = Jx + \tilde{f}(t, x), \quad \tilde{f}(t, x) = \tilde{f}(t + T, x)$$
(7.3)

$$J = \begin{bmatrix} \mu_1 & \omega_1 & 0 & 0 \\ -\omega_1 & \mu_1 & 0 & 0 \\ 0 & 0 & \mu_2 & \omega_2 \\ 0 & 0 & -\omega_2 & \mu_2 \end{bmatrix}.$$

The variables of the system can be separated into stable and critical ones

$$\dot{x}_s = J_s x_s + f_s(t, x_s, x_c)$$
(7.4a)
$$\dot{x}_c = J_c x_c + f_c(t, x_s, x_c).$$
(7.4b)

The stability and bifurcations of the system are determined from its behavior on the center manifold

$$x_s = h(t, x_c),$$
(7.5)

which can be approximated to arbitrary order as a polynomial

$$x_{i_s} = h_{i_s}(t, x_c) = H_{2i_s j_c k_c}(t) x_{j_c} x_{k_c} + H_{3i_s j_c k_c l_c}(t) x_{j_c} x_{k_c} x_{l_c} + \dots .$$
(7.6)

As in the case with constant coefficients the approximation of the center manifold starts with quadratic terms since it is tangential to the critical eigenspace of the linearized system. In contrast to the constant coefficient case the $H_{m i_s \dots}(t)$ depend on time. In (7.6) the indices with subscript s belong to the stable variables the ones with subscript c belong to the critical ones. Furthermore we use index notation and take the sum over double indices. The time dependent coefficients of $h_{i_s}(t, x_c)$ can now be calculated by differentiation of (7.5) with respect to time

$$\dot{x}_s = \frac{\partial}{\partial t} h(t, x_c) + \nabla h(t, x_c) \dot{x}_c$$
(7.7)

and substitution of (7.4a) which finally yields

$$\frac{\partial}{\partial t} \boldsymbol{h}(t, \boldsymbol{x}_c) + \nabla \boldsymbol{h}(t, \boldsymbol{x}_c)[\boldsymbol{J}_c \boldsymbol{x}_c + \boldsymbol{f}_c(t, \boldsymbol{h}(t, \boldsymbol{x}_c), \boldsymbol{x}_c)]$$
$$= \boldsymbol{J}_s \boldsymbol{h}(t, \boldsymbol{x}_c) + \boldsymbol{f}_s(t, \boldsymbol{h}(t, \boldsymbol{x}_c), \boldsymbol{x}_c). \qquad (7.8)$$

By comparison of coefficients in 7.8 we obtain the governing equations for the coefficients $H_{mi_s\ldots}(t)$ from (7.6). For each order these are linear inhomogeneous differential equations with constant coefficients which can be solved recursively. If $\boldsymbol{J}_s$ and $\boldsymbol{J}_c$ have been diagonalized the equations have the form

$$\dot{H}_{mi_s\ldots} = (\lambda_{i_s} - m_1 \lambda_{1_c} + \ldots + m_n \lambda_{n_c}) H_{mi_s\ldots} + f(t) \qquad (7.9)$$

where $f(t)$ are periodic terms. Since $\mathrm{Re}(\lambda_{i_s} - m_1 \lambda_{1_c} + \ldots + m_n \lambda_{n_c}) < 0$ the homogeneous solution vanishes such that the initial conditions can be chosen arbitrarily.

In most applications one is interested in the nonlinear stability behavior depending on parameters. In the vicinity of the stability boundary parameters can be included into the above reasoning via a state augmentation, sometimes called 'suspension trick'. A numerical example for the system under consideration is

$$\dot{p} = 0 \qquad (7.10a)$$

$$\begin{bmatrix} \dot{x}_{c1} \\ \dot{x}_{c2} \\ \dot{x}_{s1} \\ \dot{x}_{s2} \end{bmatrix} = \begin{bmatrix} p & 1 & 0 & 0 \\ -1 & p & 0 & 0 \\ 0 & 0 & -0.2 & 2 \\ 0 & 0 & -2 & -0.2 \end{bmatrix} \begin{bmatrix} x_{c1} \\ x_{c2} \\ x_{s1} \\ x_{s2} \end{bmatrix} + \begin{bmatrix} -\cos^2(t)x_{s1}^3 \\ -\cos^2(t)x_{c2}^3 \\ \cos^2(t)x_{c1}x_{c2} \\ 0 \end{bmatrix}. \qquad (7.10b)$$

In Fig. 7.1, the approximation of the center manifold up to third order is compared to the results from numerical integration for $\bar{p} = 0.005$, which can be seen to be in good agreement. For larger systems, the reduction and further manipulation of the equations can be cumbersome. Therefore, in the following we use POINCARE maps, which reduce the system to a point map [85].

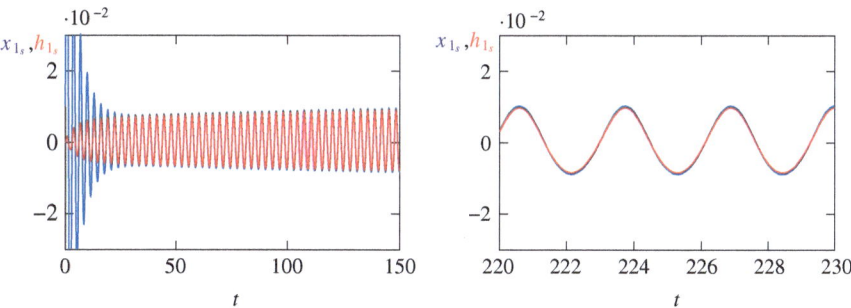

Fig. 7.1 Comparison of the approximated center manifold with numerical integration

7.2 Time Discrete Analysis by Poincare Map

Making use of the fact that the system is time-periodic, the stability and bifurcation behavior of the system (7.2) is studied through it's POINCARE map over its period. As described in [85], the POINCARE map can be expanded into a series

$$P(x_0 + \xi) = x(x_0, T) + \frac{\partial x}{\partial x_0}(x_0, T)\xi + \frac{1}{2}\frac{\partial^2 x}{\partial x_0^2}(x_0, T)\xi^2 + \ldots \qquad (7.11)$$

about the stability boundary. Since our equations are homogeneous, we are interested in the stability behavior of the trivial solution $q(t) = 0$, which is the fixed point $x_0 = 0$ of the POINCARE map. A distortion in the initial conditions ξ evolves according to the difference equation

$$\xi_{t+1} = \frac{\partial x}{\partial x_0}(0, T)\xi_t + \frac{1}{2}\frac{\partial^2 x}{\partial x_0^2}(0, T)\xi_t^2 + \ldots . \qquad (7.12)$$

We remind the reader of the intermediate steps. The parameters around the stability boundary are taken into account via a state augmentation,

$$\frac{d}{dt}\begin{bmatrix} p \\ q \\ \dot{q} \end{bmatrix} = \begin{bmatrix} 0 \\ \dot{q} \\ -M^{-1}[\Delta D(t)\dot{q} + (K + \Delta K(t))q + f(q, t)] \end{bmatrix} \qquad (7.13)$$

and in the sequel equation (7.13) is simply written as $\dot{x} = F(x, t)$. In order to calculate the derivatives of $x(x_0, T)$, Eq. (7.13) is differentiated with respect to x_0, i.e.

$$\frac{d}{dt}\frac{\partial x}{\partial x_0} = \frac{\partial F}{\partial x}\frac{\partial x}{\partial x_0} \qquad (7.14)$$

and integrated from 0 to 2π. The higher derivatives follow analogously by differentiating (7.13) with respect to x_0 repeatedly and using the results from previous computations.

Having computed the truncated expansion of the POINCARE map, our next goal is to simplify the qualitative analysis as far as possible, using normal form theory. It is known that around the stability boundary of the linear system, the dynamics of the system is determined by its behavior on the center manifold, which can be simplified further by transforming it to normal form. However, for practical reasons, we will apply normal form theory to the complete system (7.12), which implies a decoupling of the center manifold, as will be seen in the sequel. It turns out that this fact is completely analogous to the continuous case previously investigated in [23], and is used in a slightly different way in a projection method in [37] for the derivation of the generic codimension 2 bifurcations. For the reduction process, in a first step we transform (7.12) such that the linear part is in JORDAN normal form

$$x_{t+1} = \mathbf{\Lambda} x_t + f(x_t), \qquad f(x_t) = F_2 x_t^2 + F_3 x_t^3 + \ldots, \qquad (7.15)$$

where $\mathbf{\Lambda} = \mathrm{diag}(\rho_1, \ldots, \rho_n)$ and ρ_j are the FLOQUET multipliers. We divide the states into critical and non-critical variables

$$x_{t+1c} = \mathbf{\Lambda}_c x_{tc} + f_c(x_{tc}, x_{ts}), \quad |\mathbf{\Lambda}_c| = 0$$
$$x_{t+1s} = \mathbf{\Lambda}_s x_{ts} + f_s(x_{tc}, x_{ts}), \quad |\mathbf{\Lambda}_s| < 0$$

and introduce the near identity transformation

$$x_t = y_t + g(y_t), \qquad\qquad g(y_t) = G_2 y_t^2 + G_3 y_t^3 + \ldots \qquad (7.16)$$

in order to obtain

$$y_{t+1} = \mathbf{\Lambda} y_t + h(y_t), \qquad\qquad h(y_t) = H_2 y_t^2 + H_3 y_t^3 + \ldots \qquad (7.17)$$

which should be as simple as possible. Inserting (7.16) and (7.17) in (7.15) yields

$$x_{t+1} = y_{t+1} + g(y_{t+1}) = \mathbf{\Lambda} y_t + h(y_t) + g(\mathbf{\Lambda} y_t + h(y_t)) =$$
$$\mathbf{\Lambda}(y_t + g(y_t)) + f(y_t + g(y_t)). \qquad (7.18)$$

In order to solve for the unknown coefficients of $g(y_t)$ we collect terms of each order in (7.18)

$$\mathrm{Ord.0}: \ \mathbf{\Lambda} y_t \ = \mathbf{\Lambda} y_t$$
$$\mathrm{Ord.1}: \ H_2 y_t^2 = F_2 y_t^2 + \mathbf{\Lambda} G_2 y_t^2 - G_2 (\mathbf{\Lambda} y_t)^2$$
$$\vdots$$

The goal is to eliminate as many terms of the $h(y_t)$ as possible. The equations for the zero order are trivially satisfied. For each coefficient H_{2jkl} of H_2 we have

$$H_{2jkl} = \rho_j G_{2jkl} + F_{2jkl} - \rho_k \rho_l G_{2jkl} \qquad (7.19)$$

which means we can achieve $H_{2jkl} = 0$ by choosing

$$G_{2jkl} = \frac{F_{2jkl}}{\rho_j - \rho_k \rho_l} \qquad (7.20)$$

unless $\rho_j = \rho_k \rho_l$. Recursively, continuing this process, we observe that for the mth order we can eliminate all terms $H_{mm_1\ldots m_n}$ unless

$$\rho_j = \rho_1^{m_1} \ldots \rho_n^{m_n}, \qquad \sum m_i = m, \qquad (7.21)$$

which is called the resonance condition. Writing $\rho_j = e^{\lambda_j}$ with $\lambda_j = \delta_j + i\omega_j$ and taking the logarithm of (7.21), the resonance condition reads

$$\lambda_j = m_1\lambda_1 + \ldots + m_n\lambda_n, \tag{7.22}$$

which coincides with the one for the time continuous case [23]. Since for the critical variables $\delta_j = 0$, whereas for the stable variables $\delta_j < 0$, we see from (7.22) that the critical variables can only be resonant with other critical variables. Therefore the governing equations for the normal form decouple up to the order to which the reduction process is performed. A bifurcation analysis of (7.13) can be performed on the system

$$y_{t+1c} = \mathbf{\Lambda}_c y_{tc} + h_c(y_{tc}), \tag{7.23}$$

which contains only the critical variables including the parameters which were introduced by state augmentation in (7.13).

7.3 Discussion of the Neimark-Sacker Bifurcation

Depending on the type of bifurcation through which the trivial solution of (7.13) looses its stability, further information can be obtained from the resonance condition (7.21). Since the brake investigated in this paper looses stability by a NEIMARK-SACKER [37] bifurcation, the discrete version of the HOPF bifurcation, we concentrate on this case in the sequel. On the stability boundary of the trivial solution of (7.15), the critical variables correspond to a pair of complex conjugate eigenvalues $\rho_c = e^{\pm i\omega_j}$ and further critical eigenvalues $\rho_c = 1$, introduced by the state augmentation for the parameters.

Taking into account the fact that no nonlinearities occur in the equations corresponding to the parameters in (7.15), and assuming that ω_j is not a multiple of 2π, we see that only odd nonlinearities remain in the normal form (7.23), such that for the critical variables we have

$$y_{t+1c} = (\rho + a_1(\boldsymbol{p}))y_{tc} + a_3(\boldsymbol{p})y_{tc}^3 + a_5(\boldsymbol{p})y_{tc}^5 + \ldots, \tag{7.24}$$

$$\overline{y}_{t+1c} = (\overline{\rho} + \overline{a}_1(\boldsymbol{p}))\overline{y}_{tc} + \overline{a}_3(\boldsymbol{p})\overline{y}_{tc}^3 + \overline{a}_5(\boldsymbol{p})\overline{y}_{tc}^5 + \ldots, \tag{7.25}$$

where variables and coefficients are complex conjugate, since the variables in (7.15) are real. The amplitudes of limit cycles occurring after a destabilization of the trivial solution of (7.13) only depend on the norm of y_t. Therefore, for the analysis of the bifurcation behavior, it is sufficient to study only the development of

$$r_{t+1} = y_{t+1c}\overline{y}_{t+1c} = b_1(\boldsymbol{p})r_t + b_2(\boldsymbol{p})r_t^2 + b_3(\boldsymbol{p})r_t^3 + \ldots, \tag{7.26}$$

where $\sqrt{r}$ is the norm of y_{t_c}. Limit cycles and equilibrium solutions correspond to fixed points of (7.26) and can be calculated from

$$r_{t+1} - r_t = 0, \tag{7.27}$$

where only positive real roots r^* correspond to a solution. The stability of these solutions can be studied by the theory of first approximation by setting $r = r^* + \Delta r$ in (7.26) and investigating the stability of $\Delta r = 0$, which yields

$$\Delta r_{t+1} = c_1 \, \Delta r_t + c_2 \, \Delta r_t^2 + \dots. \tag{7.28}$$

In case $|c_1| < 1$, r^* is stable, in case $|c_1| > 1$, r^* is unstable.

7.4 Estimation of Basins of Attraction Through Liapounov Functions

In the previous sections we have shown that the linear stability boundary of the trivial solution is enlarged in the parameter space when the eigenvalues of the unperturbed problem are well separated. This does however not tell anything about the size of the basin of attraction of the trivial solution in the state space. It is however the size of the basin of attraction which decides on the practical relevance of the results obtained. Therefore, in the following we study the basin of attraction of the trivial solution using LIAPOUNOV functions, for which the following theorem holds:

Theorem: Let $V(x)$ be a LIAPOUNOV function for the time discrete system [86]

$$x_{t+1} = F(x_t) \tag{7.29}$$

with $F(0) = 0$. The bounded domain

$$G = \{x \in \mathbb{R}^n, 0 < V(x) < c, V(x_{t+1}) - V(x_t) < 0\}, \tag{7.30}$$

where c is a positive real constant, belongs to the basin of attraction of $x = 0$.
For the proof we note that the conditions on G certify that any solution vector starting in G strictly monotonically approaches the origin, as can be easily visualized by geometric intuition.

In the general case, the theorem gives only a one-sided estimate and the basin of attraction can be larger than G. However, in the case of a one dimensional system, the basin of attraction, namely the basin of attraction of the trivial solution of the POINCARE map (7.12), exactly coincides with G. In a non-degenerate setting, the resonance conditions will not be met between the stable variables, and after the normal form reduction the system can be decoupled into systems of degree one. Also, for the critical variables, stability is decided by only one scalar equation (7.28).

Therefore the estimation of the basins of attraction is exact up to the order of which the normal form reduction has been performed. We choose

$$V = \bar{y}^T y = \bar{y}_c^T y_c + \bar{y}_s^T y_s \qquad (7.31)$$

as a LIAPOUNOV function, with which according to the previous theorem, the exact stability boundary in physical coordinates can be found because of the decoupling of the y_i. Following HOCHLENERT [23] we can transform back the limits of the basins of attraction to the physical coordinates. To this end, we have to invert the near identity transform (7.16). Since the transformation (7.16) is continuously differentiable, by the implicit function theorem it follows that it is invertible for sufficiently small x and y can be expanded in terms of x

$$y_t = x_t + k(x_t), \qquad k(x_t) = K_2 x_t^2 + K_3 x_t^3 + \ldots \qquad (7.32)$$

The coefficients K_i of the expansion in (7.32) can be calculated analogously to the ones of G_i in (7.16). Finally, the transformation to physical coordinates reads

$$x_t = \Phi \xi_t, \qquad (7.33)$$

where Φ is the matrix of eigenvectors corresponding to the linear part of (7.11). In physical coordinates the LIAPOUNOV function now reads

$$V_t = \bar{\xi}_t^T \bar{\Phi}^T \Phi \xi_t + o(\xi_t^2), \qquad (7.34)$$

where $\bar{\Phi}^T \Phi$ is clearly positive definite and $o(\xi_t^2)$ are terms of cubic or higher order. Using (7.12) V_{t+1} can easily be calculated. The basin of attraction of the trivial solution of (7.12) can now be estimated in the physical coordinates of the point mapping (7.12) using the theorem mentioned above. The basins of attraction are exact up to the order that the normal form reduction has been carried out. It is clear that the estimate is conservative and also higher order nonlinearities in (7.12) are taken care of. The procedure only makes sense if the normal form reduction is invertible.

From the investigation it is clear that the estimate of the basin of attraction of the trivial solution gets larger with decreasing $|\rho|$. At the same time the usage of LIAPOUNOV functions assures that possibly existing other attractors, which may have been overlooked in the critical variables are taken into consideration.

7.5 Application to a Disk Brake Model

As an example we now investigate the disk brake model from Sect. 4.3. A picture of the wobbling disk model can be found in Fig. 4.4 the modeling assumptions are stated in Sect. 4.3 where the stability behavior of the linearized equations of motion was investigated. A linear stability analysis can explain the excitation mechanism for brake squeal however in the case of instability infinite vibration amplitudes are encountered which is of course not realistic. Furthermore from the linearized stability analysis it is not clear for what range about the trivial solution it is valid. In practice in case of instabilities limit cycle amplitudes are determined by nonlinearities and also the range of validity of the stability analysis. For brake models with symmetric rotors investigations in this direction have been performed in [22]. In order to account for the nonlinearities in the asymmetric brake model, the Eq. (4.16) are augmented by material nonlinearities originating from the pad material. Taking into account that the amplitudes of a squealing brake are very small, it is assumed that the stiffness characteristic of the pad is dominant, so that geometric nonlinearities can be neglected. The nonlinear equations of motion therefore read

$$M\ddot{q} + \Delta D(t)\dot{q} + (K + \Delta K(t))q + f(q,t) = 0, \tag{7.35a}$$

$$f_j(q,t) = k_3(F_{kj})^3 + k_5(F_{kj})^5, \tag{7.35b}$$

where F_{kj} is the contribution of the geometrically linear elastic restoring terms of the pad to the equations of motion. Two stiffness characteristics that have been used in the calculations are shown in Fig. 7.2.

In this section, we calculate the bifurcation equations of the brake model up to fifth order with respect to the mechanical degrees of freedom. Whereas the 'suspension trick' is probably the easiest way to understand the dependence of the bifurcation equations on the parameters, in our application it turns out to be practically infeasible. The reason is that the dependence of the solution of the differential equation (7.35) on the parameters is in fact not polynomial and the corresponding expansion converges very slowly, so that low order approximations are only valid in a very small neighborhood of the origin. This can be easily understood by considering the analytical example

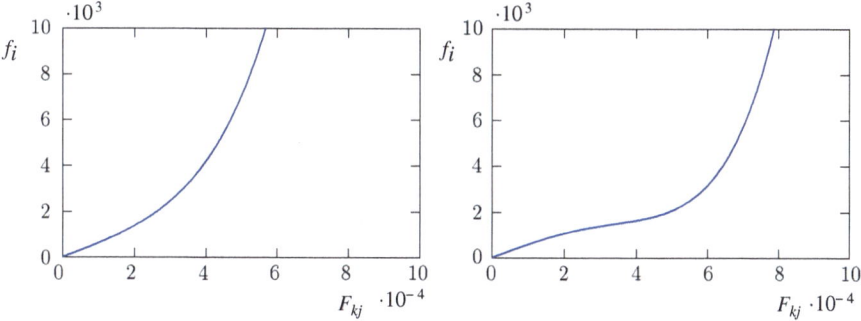

Fig. 7.2 Stiffness characteristic for $k_3 = 20 \cdot 10^{12}$ 1/(Nm)^2, $k_5 = 50 \cdot 10^{18}$ 1/(Nm)^4 (*left*) and $k_3 = -20 \cdot 10^{12}$ 1/(Nm)^2, $k_5 = 50 \cdot 10^{18}$ 1/(Nm)^4 (*right*)

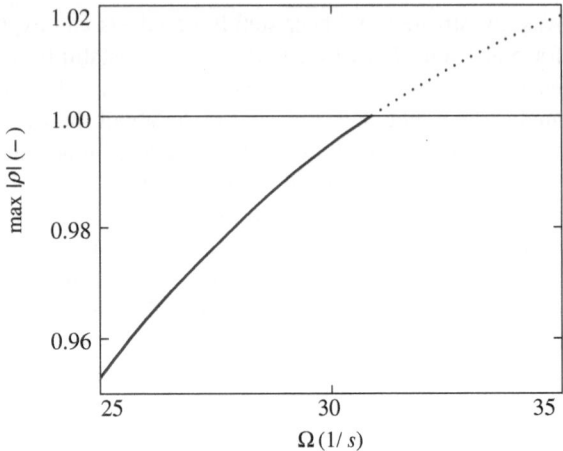

Fig. 7.3 Dependence of max $|\rho|$ on Ω for $\kappa_t = 1\,\%_o$

$$\dot{p} = 0, \tag{7.36a}$$

$$\dot{x} = -p\,k\,\sin(t)\,x, \tag{7.36b}$$

of which the solution $x = e^{k\,p\,\cos(t)}$ can easily be found by variation of the constant. Since the dependence of the solution on the parameter k is exponential, the convergence of a TAYLOR series with respect to p is very slow. The same holds true for the oscillatory dependence of the solution on the parameter, which is the case for the brake model, where many cycles occur per revolution of the disk.

In order to avoid truncation errors and taking into account that the FLOQUET multipliers as well as the nonlinearities in (7.35) are not very sensitive around the critical angular velocity Ω^* (cf. Fig. 7.3), we only perform the expansion of the POINCARE map with respect to the mechanical degrees of freedom q and unfold the equation for the norm of the normal form (7.28) with respect to Ω around Ω^*

$$r_{t+1} = \left(1 + 2\,\frac{d|\rho|}{d\Omega}\bigg|_{\Omega^*}(\Omega - \Omega^*)\right) r_t + b_2\,r_t^2 + b_3\,r_t^3 + \ldots, \tag{7.37}$$

where $d|\rho|/d\Omega$ can be calculated analytically from the expansion of the POINCARE map with respect to q and Ω [68]. When dealing with (7.37) one has to truncate the series such that no orders higher that the ones considered in the POINCARE expansion are kept, since otherwise singularities can be induced into the equations by small nonlinearities that have been removed from lower orders in the normalization process. In the following, we demonstrate that the trivial solution of (7.35) can loose its stability through a super- or a subcritical Hopf bifurcation. For the nonlinear stiffness parameters

$$k_3 = 20 \cdot 10^{12}\ 1/(\text{Nm}^2)\,, \qquad\qquad k_5 = 50 \cdot 10^{18}\ 1/(\text{Nm}^4) \tag{7.38}$$

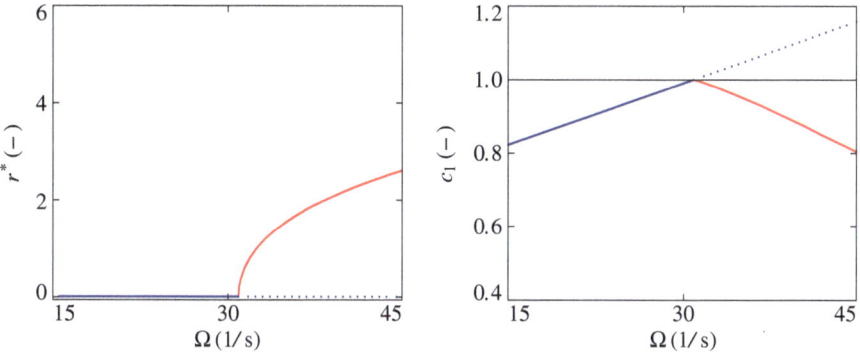

Fig. 7.4 Bifurcation diagram and stability chart (coefficient c_1 from (7.28)) for the supercritical case and $\kappa_t = 1\,\%_\circ$

we obtain the supercritical case presented in Fig. 7.4 whereas for

$$k_3 = -20 \cdot 10^{12}\ 1/(\mathrm{Nm}^2)\,, \qquad k_5 = 50 \cdot 10^{18}\ 1/(\mathrm{Nm}^4) \qquad (7.39)$$

we obtain a subcritical bifurcation as shown in Fig. 7.5.

These results coincide with the ones for the time continuous case in [23], where especially the subcritical case goes along well with observations made in the laboratory. Finally we study the dependence of the norm of the limit cycle on the rate of asymmetry in the brake rotor. In Fig. 7.6 we compare the limit cycle amplitudes for an asymmetry of $\kappa_t = 0.1,\ 1.0,\ 2.5\,\%_\circ$ for the sub- and the supercritical case. It is seen that the limit cycle amplitudes shrink with increasing asymmetry whereas the critical angular velocity rises. For asymmetries $\kappa_t > 3\,\%_\circ$ no technically relevant critical velocity can be found.

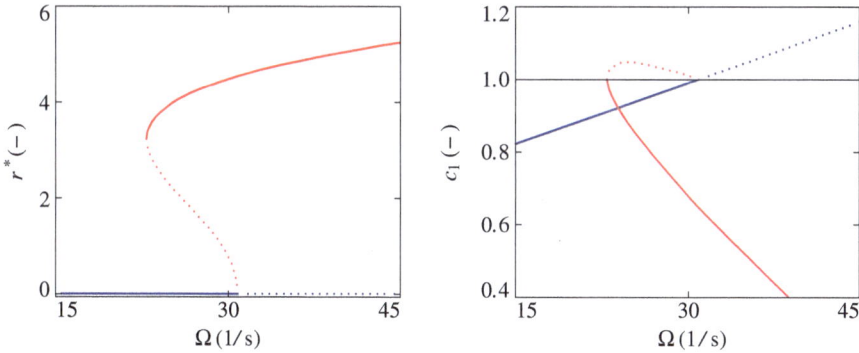

Fig. 7.5 Bifurcation diagram and stability chart (coefficient c_1 from (7.28)) for the subcritical case and $\kappa_t = 1\,\%_\circ$

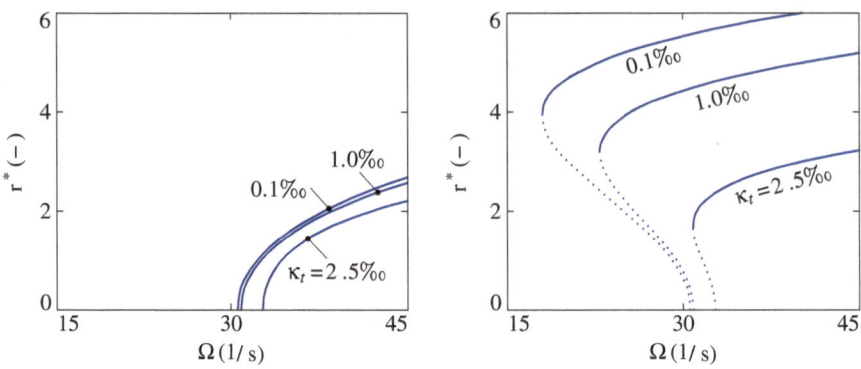

Fig. 7.6 Comparison of different asymmetries for the sub and supercritical case

Fig. 7.7 Dependence of $\max |\rho|$ on κ_t for $\Omega = 35$ 1/s

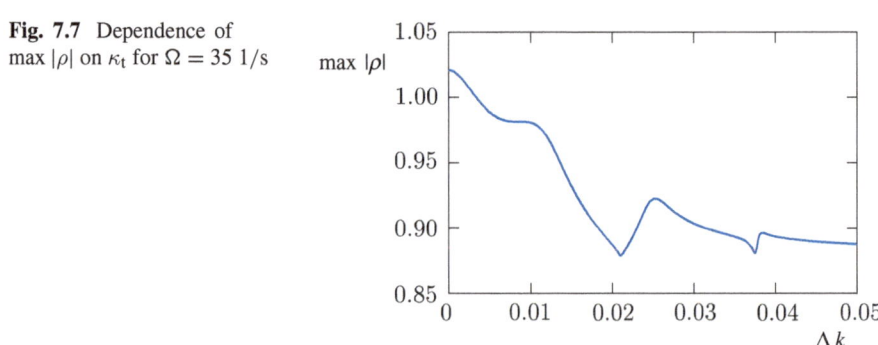

The results show that an introduction of asymmetry of the rotor not only increases the linear stability boundary but is also beneficial from a nonlinear point of view. Taking into account the results of Sect. 7.4 and taking a look at the relation of the rotor asymmetry and the largest FLOQUET multiplier, as shown in Fig. 7.7 for $\Omega = 35$ 1/s, we observe that the significant decrease of $\max |\rho|$ assures that the stabilization due to asymmetry is robust and no unstable attractor lies in a reasonable vicinity of $x = 0$.

7.6 Summary

In this chapter we have developed a method to investigate the nonlinear stability and bifurcation behavior of equations of motion with periodic coefficients. We have seen that, in analogy to the constant coefficient case, a time dependent center manifold reduction is possible. For a detailed study of the bifurcation behavior, we studied a semi-analytical approximation of the POINCARE map of the system and performed a normal form analysis. We saw that this includes a center manifold reduction. Finally we exploited how the bifurcation equations can be used to construct LIAPOUNOV functions for the estimation of basins of attraction for solutions. The results show that the approaches to split multiple frequencies of a rotor in order to prevent squeal, which were based on linear models are also meaningful in the nonlinear case.

Chapter 8
Conclusion

In this book we have studied methods for a robust design of rotors against self-excited vibrations. The occurrence of self-excited vibrations in engineering applications is often unwanted and in many cases difficult to model. Thinking of complex systems such as machines with many components and mechanical contacts, it is important to have guidelines for the design, so that the functionality is robust against small imperfections, since a detailed modeling of all relevant effects is most often impossible. Although for many phenomena sophisticated models exist or will be developed, in the future it is indispensible for the engineer to separate the relevant from the irrelevant effects and to design robust structures. For mechanical vibration problems this means that although not everything can be modeled, a way has to be found to obtain a structure which fulfills the requirements. In this book we have discussed the question on how to design a structure such that unwanted self-excited vibrations do not occur. We have shown theoretically and practically that the old design rule to avoid multiple eigenvalues is aiming in the right direction and have optimized structures accordingly. This extends results for the well known flutter problem in which equations of motion with constant coefficients occur to the case of a linear conservative system with arbitrary time periodic perturbations.

After investigating the mathematics of the underlying perturbation problem we have considered the splitting of eigenvalues as an optimization problem and shown important mathematical properties. Unfortunately the optimization problem is highly nonlinear and non-convex so that it is difficult in practice to develop an algorithm that is guaranteed to converge to the global optimum. However even the local optima obtained achieve the desired technical benefit. With a view to applications, we have considered simple models of relevant technical structures, in which avoiding multiple eigenfrequencies eliminates self-excited vibrations. This includes discrete and continuous systems. For both cases we have discussed the importance of the choice of coordinates for the mathematical structure of the equations of motion. In particular we have seen that in many cases gyroscopic terms can be eliminated by a coordinate transformation. The models discussed can be used for brake squeal, paper calenders and other applications with rotors in frictional contact or systems with moving loads.

G. Spelsberg-Korspeter, *Robust Structural Design against Self-Excited Vibrations*, SpringerBriefs in Applied Sciences and Technology, DOI: 10.1007/978-3-642-36552-2_8, © The Author(s) 2013

For the problem of brake squeal we have developed a structural optimization problem to avoid squeal by a robust design of the brake rotor. We have investigated several approaches yielding insight into the practical difficulties. In order to be able to optimize structural modifications efficiently we have introduced an approach, in which the underlying structure and the modification can be discretized separately. In experiments we validated the major conclusions of this book on a disk brake and a singing wine glass.

Finally we have investigated the nonlinear behavior of the systems under consideration using center manifold and normal form theory. It could be shown qualitatively that the goals pursued for an optimization in the linear regime are also meaningful for the nonlinear case.

The question now arises whether the conclusions from this book also solve the brake squeal problem in practice. Modifications of the brake rotor are certainly easy and cheap to realize in the manufacturing process of casting. This book in the opinion of the author reveals severe shortcomings in the numerical approaches used in industry. Although industry relies more and more on simulation of structures, for the brake squeal problem one is far from being able to accurately predict self-excited vibrations in models. For years development engineers have only considered an eigenvalue problem with constant coefficients in order to predict squeal. This results in extremely sensitive models, which yield inaccurate results concerning the excitation of brake squeal, since asymmetries of the rotor cannot be considered. The results of this book hopefully show that asymmetries of the rotor are practically relevant and that the current industrial models have to be extended to cover the relevant effects.

The next goal following from this work is to extend the optimization problem for the brake rotor to a larger design space, and more important, to incorporate further industrially relevant constraints. A first step in this direction is the consideration and investigation of thermal effects in the optimization problem. Thermal effects are irrelevant for the speed range of vehicles in which squeal occurs, however, the brake rotor has of course to be designed for heavy braking maneuvers with significant thermal loads as well. These more practical aspects are pursued in a PhD project together with an industrial partner.

References

1. Allgaier, R., Gaul, L., Keiper, W., Willner, K.: Mode lock-in and friction modelling. In: International Conference on Computational Methods in Contact Mechanics, pp. 35–47 (1999)
2. Brommundt, E.: High-frequency self-excitation in paper calenders. Technische Mechanik 29(1), 60–85 (2009)
3. Bucher, I., Braun, S.: The structural modification inverse problem: an exact solution. Mech. Syst. Signal Process. 7(3), 217–238 (1993)
4. Cullum, J., Donath, W., Wolfe, P.: The minimization of certain nondifferentiable sums of eigenvalues of symmetric matrices. Nondiffer. optim. 3, 35–55 (1975)
5. Cunefare, K., Graf, A.: Experimental active control of automotive disc brake rotor squeal using dither. J. Sound Vib. 250(4), 579–590 (2002)
6. Dávid, A., Sinha, S.: Versal deformation and local bifurcation analysis of time-periodic nonlinear systems. Nonlinear Dyn. 21(4), 317–336 (2000)
7. Fidlin, A.: Nonlinear Oscillations in Mechanical Engineering. Springer, New York, 2006
8. Fidlin, A., Drozdetskaya, O., Waltersberger, B.: On the minimal model for the low frequency wobbling instability of friction discs. Eur. J. Mech. Solids 30, 665–672 (2011)
9. Fieldhouse, J.D., Steel, W.P., Talbot, C.J., Siddiqui, M.A.: Brake noise reduction using rotor asymmetry. In: International Conference on Vehicle Braking and Chassis Control (Braking 2004), pp. 209–222 (2004)
10. Fieldhouse, J.D., Steel, W.P., Talbot, C.J., Siddiqui, M.A.: Rotor assymetry used to reduce disk brake noise. SAE Technical paper 2004–01-2797 (2004)
11. Hader, P.: Selbsterregte Schwingungen von Papierkalandern. Ph.D. thesis, RWTH Aachen (2005)
12. Hagedorn, P., von Wagner, U.: Smart pads: a new tool for the suppression of brake squeal? Fortschritts Berichte-VDI Reihe 12 Verkehrstechnik Fahrzeugtechnik, pp. 153–172 (2004)
13. Hagedorn, P.: Technische Schwingunslehre: Lineare Schwingungen kontinuierlicher mechanischer Systeme. vol. 2. Springer, Berlin (1989)
14. Hamdi, A.: A Moreau-Yosida regularization of a difference of two convex functions. Appl. Math. E-Notes 5, 164–170 (2005)
15. Hartman, P.: On functions representable as a difference of convex functions. Pac. J. Math. 9(3), 707–713 (1959)
16. Heilig, J.: Instabilitäten in Scheibenbremsen durch dynamischen Reibkontakt, Ph.D. thesis, Technische Universität Darmstadt (2002)
17. Helmberg, C.: Semidefinite programming. Eur. J. Oper. Res. 137(3), 461–482 (2002)
18. Hertz, H.: Die Prinzipien der Mechanik, Wissenschaftliche Buchgesellschaft Darmstadt (1963)
19. Hervé, B., Sinou, J., Mahé, H., Jézéquel, L.: Analysis of squeal noise and mode coupling instabilities including damping and gyroscopic effects. Eur. J. Mech. A Solids 27(2), 141–160 (2008)
20. Hiriart-Urruty, J., Lemaréchal, C.: Fundamentals of Convex Analysis. Springer, Berlin (2001)

G. Spelsberg-Korspeter, *Robust Structural Design against Self-Excited Vibrations*, SpringerBriefs in Applied Sciences and Technology, DOI: 10.1007/978-3-642-36552-2, © The Author(s) 2013

21. Hochlenert, D.: Selbsterregte Schwingungen in Scheibenbremsen: Mathematische Modellbildung und aktive Unterdrückung von Bremsenquietschen. Ph.D. thesis, Technische Universität Darmstadt (2006)
22. Hochlenert, D.: Nonlinear stability analysis of a disk brake model. Nonlinear Dyn. **58**(1), 63–73 (2009)
23. Hochlenert, D.: Dimension reduction of nonlinear dynamical systems: comparison of the center manifold theory and the method of multiple scales. GAMM Jahrestagung Karlsruhe (2010)
24. Hochlenert, D., Spelsberg-Korspeter, G., Hagedorn, P.: Friction induced vibrations in moving continua and their application to brake squeal. Trans. ASME J. Appl. Mech. **74**, 542–549 (2007)
25. Hochlenert, D., von Wagner, U., Hornig, S.: Bifurcation behavior and attractors in vehicle dynamics. Mach. Dyn. Probl. **33**(2), 57–73 (2009)
26. Hoffmann, N., Fischer, M., Allgaier, R., Gaul, L.: A minimal model for studying properties of the mode-coupling type instability in friction induced oscillations. Mech. Res. Commun. **29**, 197–205 (2002)
27. Ibrahim, R.A.: Friction-induced vibration, chatter, squeal and chaos, part I: mechanics of contact and friction. ASME Appl. Mech. Rev. **47**(7), 227–253 (1994)
28. Ilanko, S.: Existence of natural frequencies of systems with artificial restraints and their convergence in asymptotic modelling. J. Sound Vib. **255**(5), 883–898 (2002)
29. Ilanko, S.: Introducing the use of positive and negative inertial functions in asymptotic modelling. Proc. R. Soc. A Math. Phys. Eng. Sci. **461**(2060), 2545 (2005)
30. Ilanko, S., Dickinson, S.: Asymptotic modelling of rigid boundaries and connections in the Rayleigh-Ritz method. J. Sound Vib. **219**(2), 370–378 (1999)
31. Ilanko, S., Williams, F.: Wittrick-Williams algorithm proof of bracketing and convergence theorems for eigenvalues of constrained structures with positive and negative penalty parameters. Int. J. Numer. Methods Eng. **75**(1), 83–102 (2008)
32. Jearsiripongkul, T.: Squeal in floating caliper disk brakes: a mathematical model. Ph.D. thesis, Technische Universität Darmstadt (2005)
33. Karapetjan, A.V.: The stability of nonconservative systems. Vestnik Moskovskogo Universiteta Serija 1, mathematika, mechanika **30**(4), 109–113 (1975)
34. Kinkaid, N.M., O'Reilly, O.M., Papadopoulos, P.: Automotive disc brake squeal. J. Sound Vib. **267**, 105–166 (2003)
35. Kirillov, O.N., Seyranian, A.P.: Stabilization and destabilization of a circulatory system by small velocity-dependent forces. J. Sound Vib. **283**, 781–800 (2005)
36. Kröger, M., Neubauer, M., Popp, K.: Experimental investigation on the avoidance of self-excited vibrations. Philos. Trans. R. Soc. A Math. Phys. Eng. Sci. **366**(1866), 785–810 (2008)
37. Kuznetsov, Y., Kuznetsov, I., Kuznetsov, Y.: Elements of Applied Bifurcation Theory. Springer, New York (1998)
38. Kwak, M.K., Han, S.: Free vibration analysis of rectangular plate with a hole by means of independent coordinate coupling method. J. Sound Vib. **306**, 12–30 (2007)
39. Kyprianou, A., Mottershead, J., Ouyang, H.: Assignment of natural frequencies by an added mass and one or more springs. Mech. Syst. Signal Process. **18**(2), 263–289 (2004)
40. Lakhadanov, V.M.: On stabilization of potential systems. PMM **39**(1), 53–58 (1975)
41. Lancaster, P., Tismenetsky, M.: The Theory of Matrices, 2 edn. Academic Press Inc., New York (1985)
42. Lang, A., Newcomb, T.: An experimental investigation into drum brake squeal. In: Proceedings of EAEC Conference on New Developments in Powertrain and Chassis Engineering, Strasbourg, pp. 431–444 (1989)
43. Lewis, A.: The mathematics of eigenvalue optimization. Math. Program. **97**(1), 155–176 (2003)
44. Lewis, A.; Overton, M.: Eigenvalue optimization. Acta Numerica **5**, 149–190, (1996)

45. Martinez-Legaz, J., Seeger, A.: A formula on the approximate subdifferential of the difference of convex functions. Bull. Aust. Math. Soc. **45**(01), 37–41 (1992)
46. Massi, F., Berthier, Y., Akay, A.: On the need for new approaches for brake squeal prediction and suppression. In: 6th European Conference on Braking JEF, Lille (2010)
47. Millner, N.: An analysis of disc brake squeal. SAE Paper 780332, SAE Congress Detroit (1978)
48. Müller, P.C.: Stabilität und Matrizen. Springer, Berlin (1977)
49. Müller, P., Schiehlen, W.: Lineare Schwingungen. Akad. Verl.-Anst. (1976)
50. Mottershead, J.: Vibration- and friction-induced instability in disks. Shock Vib. Dig. **30**(1), 14–31 (1998)
51. Mottershead, J., Lallement, G.: Vibration nodes, and the cancellation of poles and zeros by unit-rank modifications to structures. J. Sound Vib. **222**(5), 833–851 (1999)
52. Mottershead, J., Mares, C., Friswell, M.: An inverse method for the assignment of vibration nodes. MSSP- Mech. Syst.Signal Process. **15**(1), 87–100 (2001)
53. Neubauer, M., Oleskiewicz, R.: Suppression of brake squeal using shunted piezoceramics. J. Vib. Acoust. **130**, 021005 (2008)
54. Nishiwaki, M., Harada, H., Okamura, H.: Study on disc brake squeal. SAE Technical paper (1989)
55. North, M.R.: Disc brake squeal—a theoretical model. Technical report, Motor Industry Research Assosiacion (1972)
56. North, M.: Disc brake squeal. Proc. IMechE **C38**, 169–176 (1976)
57. Ouyang, H.: Prediction and assignment of latent roots of damped asymmetric systems by structural modifications. Mech. Sys. Signal Process. **23**(6), 1920–1930 (2009)
58. Ouyang, H., Mottershead, J.E.: Dynamic instability of an elastic disk under the action of a rotating friction couple. J. Appl. Mech. **71**, 753–758 (2004)
59. Ouyang, H., Nack, W., Yuan, Y., Chen, F.: Numerical analysis of automotive disc brake squeal: a review. Int. J. Veh. Noise Vib. **1**(3), 207–231 (2005)
60. Overton, M.: Large-Scale Optimization of eigenvalues. Citeseer (1990)
61. Overton, M., Womersley, R.: Optimality conditions and duality theory for minimizing sums of the largest eigenvalues of symmetric matrices. Math. Program. **62**(1), 321–357 (1993)
62. Overton, M., Womersley, R.: On the Sum of the Largest Eigenvalues of a Symmetric Matrix. New York University, Department of Computer Science, Courant Institute of Mathematical Sciences (1991)
63. Popp, K., Rudolph, M., Kroger, M., Lindner, M.: Mechanisms to generate and to avoid friction induced vibrations. VDI Berichte **1736**, 1–16 (2002)
64. Serre, J.: Linear Representations of Finite Groups, vol. 42. Springer Verlag, Berlin (1977)
65. Seyranian, A.P.: Resonance domains for the hill equation. Doklady Phys. **46**(1), 41–44 (2001)
66. Seyranian, A.P., Kirillov, O.N., Mailybaev, A.A.: Coupling of eingenvalues of complex matrices at diabolic and exceptional points. J. Phys. A Math. Gen. **38**, 1723–1740 (2005)
67. Seyranian, A.P., Solem, F., Petersen, P.: Stability analysis for multi-parameter linear periodic systems. Arch. Appl. Mech. **69**, 160–180 (1999)
68. Seyranian, A., Privalova, O.: The Lagrange problem on an optimal column: old and new results. SStruct. Multidiscip. Optim. **25**(5), 393–410 (2003)
69. Seyranian, A., Solem, F., Pedersen, P.: Multi-parameter linear periodic systems: sensitivity analysis and applications. J. Sound Vib. **229**(1), 89–111 (2000)
70. Shin, K., Brennan, M., Oh, J., Harris, C.: Analysis of disc brake noise using a two-degree-of-freedom model. J. Sound Vib. **254**(5), 837–848 (2002)
71. Soedel, W.: Vibrations of Shells and Plates. Marcel Dekker, Inc, New York (1981)
72. Spelsberg-Korspeter, G.: Breaking of symmetries for stabilization of rotating continua in frictional contact. J. Sound Vib. **322**, 798–807 (2009)
73. Spelsberg-Korspeter, G.: Structural optimization for the avoidance of self-excited vibrations based on analytical models. J. Sound Vib. **329**, 4829–4840 (2010)

74. Spelsberg-Korspeter, G.: Eigenvalue optimization against brake squeal: symmetry, mathematical background and experiments. J. Sound Vib. **331**, 4259–4268 (2012)
75. Spelsberg-Korspeter, G., Hochlenert, D., Hagedorn, P.: On the structural optimization of brake disks with respect to their vibrational behavior. In: Proceedings of the 7th International Workshop on Structural Health Monitoring (2009)
76. Spelsberg-Korspeter, G.; Hochlenert, D.; Hagedorn, P.: Non-linear investigation of an asymmetric disk brake model. Proc. Inst. Mech. Eng. C J. Mech. Eng. Sci. (2011). doi: 10.1177/0954406211408531
77. Spelsberg-Korspeter, G., Hochlenert, D., Kirillov, O.N., Hagedorn, P.: In- and out-of-plane vibrations of a rotating plate with frictional contact: investigations on squeal phenomena. Trans. ASME J. Appl. Mech. **76**(4), 041006/1-14 (2009)
78. Spelsberg-Korspeter, G., Hochlenert, D., Vomstein, T., Hagedorn, P.: Combination of finite elements with analytical contact formulations for the analysis of brake squeal. VDI Berichte, vol. 2218, pp. 279–288 (2010)
79. Spelsberg-Korspeter, G., Kirillov, O., Hagedorn, P.: Modeling and stability analysis of an axially moving beam with frictional contact. J. Appl. Mech. **75**(3), 031001/1-10 (2008)
80. Spelsberg-Korspeter, G., Wagner, A.: Discretization of structures using a negative stiffness approach in the context of structural optimization. In: Proceedings of the 8th International Symposium of Continous Systems (2011)
81. Spurr, R.: A theory of brake squeal. Proc. Autom. Div. Inst. Mech. Eng. **1**, 33–40 (1961/1962)
82. Stiefel, E.: Fässler, A.: Gruppentheoretische Methoden und ihre Anwendung. BG Teubner (1979)
83. Tao, P., An, L.: A DC optimization algorithm for solving the trust-region subproblem. SIAM J. Optim. **8**(2), 476–505 (1998)
84. Toh, K., Todd, M., Tutuncu, R.: SDPT3 a Matlab software package for semidefinite programming. Optim. Methods Softw. **11**(12), 545–581 (1999)
85. Troger, H., Steindl, H.: Nonlinear Stability and Bifurcation Theory. Springer, Wien (1991)
86. Unbehauen, H.: Regelungstechnik. Vieweg, Wiesbaden (2000)
87. van Haag, R.: über die Druckverteilung und die Papierkompression im Walzenspalt eines Kalanders. Ph.D. thesis, TU Darmstadt (1993)
88. Vandenberghe, L., Boyd, S.: Semidefinite programming. SIAM Rev. **38**(1), 49–95 (1996)
89. Vishik, M.I., Lyusternik, L.A.: The asymtotic behaviour of solutions of linear differential equations with large or quickly changing coefficients and boundary conditions. Russ. Math. Surv. **3**, 23–91 (1960)
90. Vomstein, T.: Strukturmodifikation moderner Bremsscheiben: Mathematische Modellierung und passive Unterdrückung von Bremsenquietschen. Ph.D. thesis, TU Darmstadt (2007)
91. von Wagner, U., Hochlenert, D., Hagedorn, P.: Minimal models for disk brake squeal. J. Sound Vib. **302**(3), 527–539 (2007)
92. Von Wagner, U., Hochlenert, D., Jearsiripongkul, T., Hagedorn, P.: Active Control of Brake Squeal Via 'Smart Pads' (2004)
93. Wagner, A.: Untersuchungen zur Eigenwertoptimierung von dynamischen Strukturen anhand konkreter mechanischer Anwendungsbeispiele. Diplomarbeit TU Darmstadt (2010)
94. Williams, F., Ilanko, S.: The use of the reciprocals of positive and negative inertial functions in asymptotic modelling. Proc. R. Soc. A Math. Phys. Eng. Sci. **462**(2070), 1909 (2006)
95. Wooden, S., Sinha, S.: Analysis of periodic-quasiperiodic nonlinear systems via Lyapunov-Floquet transformation and normal forms. Nonlinear Dyn. **47**(1), 263–273 (2007)
96. Yakubovich, V.A., Starzhinskii, V.M.: Linear Differential Equations with Periodic Coefficients, vol. 1. Wiley, New York (1975)